Polygyny and
Sexual Selection in
Red-winged
Blackbirds

MONOGRAPHS IN
BEHAVIOR AND ECOLOGY

Edited by John R. Krebs and
Tim Clutton-Brock

Polygyny and Sexual Selection in Red-winged Blackbirds

WILLIAM A. SEARCY and
KEN YASUKAWA

Princeton University Press
Princeton, New Jersey

Library of Congress Cataloging-in-Publication Data

Searcy, William A., 1950–
Polygyny and sexual selection in red-winged blackbirds /
    William A. Searcy and Ken Yasukawa.
p.   cm. —(Monographs in behavior and ecology)
Includes bibliographical references (p.   ) and index.
ISBN 0-691-03686-1 (cl)—ISBN 0-691-03687-X (pbk.)
1. Red-winged blackbird—Behavior.   2. Sexual behavior
in animals.   3. Sexual selection in animals.
I. Yasukawa, Ken, 1949–   . II. Title. III. Series.
QL696.P2475S43   1994
598.8'81—dc20   94-19415

This book has been composed in Times Roman

Printed in the United States of America

10 9 8 7 6 5 4 3 2 1
10 9 8 7 6 5 4 3 2 1
(Pbk.)

To Margaret and Sondra

Contents

List of Figures
and Tables

TABLES

Acknowledgments

Research on red-winged blackbirds seems to have attracted a particularly cooperative, noncontentious set of workers, in stark contrast to other areas we could name within behavioral ecology. There have been disagreements and controversies in redwing work, to be sure, but they have been amicable, and the researchers involved have almost universally remained on good terms with one another. In writing this book, we have benefited greatly from the cooperative spirit pervading the field, which enabled us to ask for and receive help from all and sundry.

We would like to single out for special thanks four of the heavy hitters in redwing research: Les Beletsky, Gordon Orians, Patrick Weatherhead, and David Westneat. All four could not have been more helpful: all read and commented on the entire book, while at the same time providing access to unpublished manuscripts and data. Others who read and commented on all the chapters include L. Scott Johnson and Kristina Hannam. Scott was of particular assistance in assembling material on polygyny in other species of birds. Linda Whittingham provided help with the chapter on parental care, again in the form of both comments and unpublished results. Others who provided us with advance copies of manuscripts include Jaroslav Picman, Maynard Milks, Michelle Leptich, Karen Metz, Robert G. Clark, Dave Shutler, Robert Montgomerie, H. Lisle Gibbs, Peter Boag, Raleigh Robertson, Donald Kroodsma, Frances James, Torgier Johnsen, James Hengeveld, Christopher Rogers, James Blank, Val Nolan, Jr., Daniel Cristol, Malte Andersson, and John Wingfield. We are grateful to all.

We also thank those who have helped us with our own research on red-winged blackbirds. For W. A. S. these include Gordon Orians, for getting him started and providing continued inspiration through the years; Peter Marler, for advice and support during postdoctoral work; Margaret Hairston Searcy, Stephan Coffman, George Reese, William DeRagon, Kristin Williams, Laurel Roberts, John Mull, Mark Patterson, Christopher A. Searcy, and others for help in the field; and Richard Hartman, Dolly Smith, and Ken Huser for logistical support. K. Y. thanks John M. Emlen, for sparking his interest in behavioral ecology, and Val Nolan, Jr., for forcing him to observe redwings in the advanced ornithology course; Peter Marler, for support and encouragement during postdoctoral work; Richard Newsome, for his stewardship of Newark Road Prairie; Dave Enstrom and a small army of undergraduates, including Elyse Bick, Wolfe Wagman, Anne Todd, Judy McClure, Becky Boley, Sue Simon, Juli Zanocco, Lynn Whittenberger, Kris Agard, Scott Walter, Margaret Russell, Todd Ditman, Brian Hodgkins,

Ming-Ming Sun, Tracy Nielsen, Todd Bothel, Fran Leanza, Chris King, Kiva Wood, Greg Hess, Caitrine Hellenga, William Jones, Jennifer Nerat, Adam Robbins, Teresa Friedrich, Julie Burford, and Courtney Thomas, for help in the field; and Sondra Fox, for constant support and understanding. Finally, both of us thank the National Science Foundation for financial support.

We gratefully acknowledge the University of Chicago Press for permission to reprint Figure 1.2; Springer-Verlag for permission to reprint Figures 2.3 and 8.10; Academic Press for permission to reprint Figures 2.4, 2.5, 3.7, 3.8, 4.1, 5.8, 5.10, 8.8, 9.2, and 9.4; Sigma Xi and *American Scientist* for permission to reprint Figures 3.3, 3.4, 3.5, and 3.6; the National Research Council of Canada and the *Canadian Journal of Zoology* for permission to reprint Figures 5.1 and 9.1; the American Ornithologists' Union and *The Auk* for permission to reprint Figure 5.7; the American Association for the Advancement of Science for permission to reprint Figures 8.6 and 8.9; the Society for the Study of Evolution for permission to reprint Figures 8.13, 9.5, and 9.6; the Canadian Museum of Nature for permission to reprint Figures 8.6, 8.9, and 8.11; and the Cooper Ornithological Society for permission to reprint Figure 9.3. We also gratefully acknowledge the original authors of these figures, whose names are given in the figure captions.

Polygyny and
Sexual Selection in
Red-winged
Blackbirds

1   Introduction

The behavior of red-winged blackbirds in the wild has been studied as extensively as that of any species of bird in the world. Given the traditional prominence of birds in ethology and behavioral ecology, this means that "redwing" behavior is as well known as that of any species of any taxa. That so much attention has been paid to red-winged blackbirds is due in part to certain practical advantages they offer for field study: they occupy open habitats, where they are easy to observe, and in most locales their nests are easy to locate, so that reproductive success can be followed at least over the short term. Additionally, the species is abundant over a large area where ornithologists are also thick on the ground. The popularity of redwings for study can also be attributed in part to their having been chosen for investigation by a few early workers in the 1950s, just as the current wave of interest in evolutionary studies of behavior and ecology was getting underway. It is generally true that the more that is found out about any species, the more new questions arise, so that interest in one species can snowball; this type of "runaway" (with apologies to Sir Ronald A. Fisher) is what has happened with red-winged blackbirds. The single most important reason for the interest in red-winged blackbirds, however, is that they are polygynous, and highly polygynous at that. Polygyny, in which one male has several mates, is a rare mating system in birds. The vast majority of bird species are socially monogamous, and very few indeed of the polygynous species show the degree of polygyny found in redwings. This puts red-winged blackbirds at the center of two of the hotter areas in behavioral ecology: mating systems and sexual selection.

Mating systems are categories of breeding-season social organization, and as such they sum up a great deal of the behavior and ecology of a species. The study of polygyny, especially polygyny in birds, has been central to the development of mating-systems theory (Wittenberger 1981), a body of theory that attempts to explain the evolution of mating systems within species, and their distribution across species. As the best-known of polygynous birds (the only possible rival being the pied flycatcher), it is natural to use red-winged blackbirds to test theories about the evolution of polygyny; one of our principal objectives in this book will be to use redwings for this purpose. We recognize that it is unlikely that a single hypothesis will explain the occurrence of polygyny in all species with this mating system (Searcy and

Yasukawa 1989), so theories need not stand or fall solely on how well they fit red-winged blackbirds. Still, it is logical to test polygyny theories first on those species for which the most complete data are available.

A second objective will be to explore the effects of the mating system, acting through sexual selection, on other aspects of the species' ecology and behavior. Sexual selection exercises a peculiar fascination for contemporary biologists, perhaps because sexual selection can account for the occurrence of so many traits that seem in some ways nonadaptive, such as bizarre ornaments and display behaviors. As Darwin (1871) was the first to point out, sexual selection acts more strongly in polygynous than monogamous species, making any polygynous species a natural laboratory for the study of the effects of sexual selection. It will be our contention that in red-winged blackbirds, the polygynous mating system has a pervasive influence on the biology of the species, in large part because polygyny enhances sexual selection, but for other reasons as well.

In this introductory chapter we will first define the various categories of mating systems found in birds. Second, we will describe the natural history of red-winged blackbirds, in sufficient detail to justify their assignment to the category of polygyny. Third, we will give a historical development of some of the principal hypotheses on the evolution of polygyny, which will be tested against the redwing data later in the book. Finally, we will introduce some of the concepts from the theory of sexual selection that we will employ later.

Mating Systems

There are two types of definitions for mating systems: social and genetic. Social definitions stress the social relationships that form between males and females breeding together: how many associations each individual forms simultaneously or within a season, how long associations last, and what kinds of cooperative behavior each sex performs, especially in terms of parental care. Genetic definitions stress the transmission of genes to offspring: the relative numbers of males and females transmitting genes in one season, or the number of members of the opposite sex with which each individual shares parentage within a season. Previously it was thought that the two sets of definitions were congruent, so that (for example) a male and female that formed a monogamous social relationship would also share gametes only with each other. It is interesting to consider why we so readily assumed that sexual relations would be more straightforward in animals such as birds than they are in humans, but at any rate it has by now been shown that they are not. Instead, social relationships do not always reveal how gametes are shared, in humans and nonhumans alike, so social and genetic definitions of mating systems must be kept separate.

The social definitions we will use are modified from Wittenberger (1979,

1981). Polygyny is defined as the prolonged association and mating relationship between one male and two or more females at a time. The only modification from Wittenberger here is that we do not specify an "essentially exclusive" mating relationship, which recent genetic evidence indicates may be rather rare. The other major systems are monogamy, the prolonged association and mating relationship between one male and one female at a time; polyandry, the prolonged association and mating relationship between one female and two or more males at a time; and promiscuity, in which associations between males and females are always brief. Finally, in rare cases there can be a prolonged association and mating relationship between two or more males and two or more females; such a system is called polygynandry.

These definitions stress social relationships and therefore have implications for social behavior such as intrasexual aggression, pair bonding, and parental care. They are stated so as to apply to individual mating associations rather than to populations or species. Thus one might find an occasional polygynous association (say one male and two females) in a population in which most associations are monogamous; this of course is actually the typical pattern in most "monogamous" species. By a convention proposed by Verner and Willson (1966), a population is termed polygynous if 5% or more of its males form polygynous associations. This rather lax criterion would seem to skew the classification toward polygyny, but even so over 90% of bird species still qualify as monogamous (Verner and Willson 1966, Lack 1968).

Wiley (1974) proposed genetic definitions of mating systems that hinge on the relative numbers of males and females contributing gametes to offspring within a breeding system. Polygyny, for example, is defined as a system in which more females than males contribute gametes. These definitions apply only to populations and not to individual associations. An individual-level, genetic definition for polygyny is a system in which one male contributes gametes to zygotes of more than one female within a season, whereas each female shares gametes with only one male, at least within a single breeding attempt. In this classification, monogamy means that one male and one female share gametes only with each other; and polyandry means that one female uses gametes from more than one male, whereas each male shares gametes with only one female. Promiscuity has no place in this classification. According to genetic definitions, polygynandry occurs when both males and females share gametes with multiple members of the opposite sex.

The Mating System of Red-winged Blackbirds

Red-winged blackbirds are members of the subfamily Icterinae, the New World blackbirds, in the family Emberizidae. The species' common name describes the appearance of adult males, which are a glossy black except for a red "epaulet," bordered in yellow, at the bend (wrist) of the wing. Males attain this "adult" plumage in the prebasic molt in August of their second year. Prior to this molt, one-year-old males are much browner overall, with smaller epaulets that are orange or red-orange rather than red. One-year-old males are also slightly smaller than "adult" (i.e., older) males; for example, wing lengths of one-year-olds and adults averaged 124 and 131 mm, respectively, in a Washington State population, and 121 and 128 mm in an Indiana population. We will refer to one-year-old males as "subadults," though this term is somewhat misleading in that such males actually seem to be sexually competent (Wright and Wright 1944, Payne 1969). Adult females are even smaller than subadult males, averaging 107 mm in wing length in Washington and 102 mm in Indiana. Females attain adult plumage, more or less, in their first prebasic molt in August of their natal year. Females in the adult plumage are dark brown above and brown streaked with white below. Females lack the conspicuous epaulets of adult males, but many do have a splotch of orange or red, of variable size and intensity, in the same position.

The preferred breeding habitat of red-winged blackbirds is marsh; this is where breeding densities and reproductive success tend to be highest. In arid regions, especially in western North America, redwings nest almost exclusively in marshes, but in wetter areas in the Midwest and East they breed in grassland and old fields as well. In northern populations, both sexes migrate south for the winter, and males return north to the breeding areas in advance of the females. Immediately upon return, males establish individual territories if they can, but there is always a surplus of males that are apparently unable to find positions in appropriate habitat. Subadult males are greatly over-represented in this surplus population, but there are many adult male "floaters" as well. In some cases, adult males remain as nonterritorial floaters throughout their lives.

Adult females begin arriving in northern breeding habitats some weeks after the first males. Early-arriving females usually delay a few days or weeks before nesting, whereas later arrivals may begin nesting almost immediately. Once a female settles on a territory, she is committed to a residence of about six to seven weeks if her nesting attempt is successful: approximately three or four days to build the nest, three to five days to lay eggs, 11 or 12 days to incubate, 11 days with young in the nest, and perhaps another two weeks feeding fledglings on the territory. Second-nesting ("double clutching") in the same season is rare among females who succeed in raising young on their first attempt, at least in northern marshes. Unsuccessful females usually renest at least once, and most renesting occurs on the original territory, but some females do switch territories between attempts. Females give parental care by building nests, incubating eggs, brooding young, feeding nestlings and fledglings, and defending eggs and young against predators. Males do not build, incubate, or brood, they may or may not feed nestlings and fledglings, and they perform varying amounts of defense against predators. We know of no cases in which two or more males shared a territory or in which nestlings or fledglings were fed by more than one male, although it is common for several males from neighboring territories to "mob" a potential predator together.

The average female red-winged blackbird, then, remains on a territory, associating with the male owner, for several weeks. This period qualifies as a prolonged association in avian terms, thus satisfying part of our social definition of polygyny. Furthermore, the relationship does involve mating; observations show that the majority of females copulate most often with the owners of their nesting territories (see below). The final criterion in our social definition of polygyny is that males form multiple associations simultaneously, and of course this is the key feature of the redwing mating system.

The females nesting on the territory of an individual male are referred to as his "harem," a term with unnecessary and unintended connotations but still a convenient one. Harem sizes are calculated in various ways, none of which are ideal. In Box 1.1 we demonstrate the four most commonly used estimates of harem size and discuss their relative merits. Table 1.1 shows mean and maximum harem sizes for various populations of red-winged blackbirds, along with the method of estimation used. Mean harem sizes range from 1.7 up to 6.2, and the overall maximum recorded harem size is 15. In all these populations, the great majority of male-female associations are polygynous. For example, the population in Table 1.1 with the lowest mean harem size (1.7) is a mixed marsh-upland population in New York studied by Westneat (1993a). In that population, 50 of 72 nesting females (69%) were involved in polygynous associations, as were 21 of 43 mated males (49%). In the Searcy and Yasukawa (1983) study of a marsh population in eastern Washington, 489 of 492 females (99%) bred in polygynous associations, as did 96 of 107 territorial males (90%). Other populations mostly fall within this range. Clearly, social polygyny is the norm in this species.

We can compare the degree of polygyny found in red-winged blackbirds to that found in other socially polygynous species. In many such species, maximum harem size is only two, and mean harem sizes for territorial males are approximately one (see Table 1.2). In these species, most male-female associations are not polygynous. For example, in indigo buntings only 21% of the females and 10% of the males mated in polygynous associations (Carey and Nolan 1979). Figures for marsh wrens are 46% of females and 25% of males mating polygynously (Verner and Engelsen 1970), and for great reed warblers 50% of females and 27% of males (Catchpole 1986). Thus, in these species the average adult does not mate polygynously. Of course, there do exist other species, besides redwings, for which polygyny is

BOX 1.1 CALCULATION OF HAREM SIZE

Below we outline some of the methods that have been used to estimate harem size and discuss their relative virtues. How one rates the methods depends in large part on how harem size is defined, and different definitions can be appropriate depending on the questions being addressed. We will employ harem size, first and foremost, as a measure of the degree of polygyny, and accordingly we define harem size as the maximum number of females simultaneously mated to a male within a season. For illustrative purposes, we use the methods to calculate harem size for the hypothetical nesting chronology shown in Figure 1.1.

Method 1

The total number of nests built on a territory (e.g., Smith 1943). For our example, this method would estimate harem size as six. Advantages of this method are that it does not require that females be marked, nor does it require complete nesting chronologies. The major disadvantage is that it will overestimate harem size (as we define it) when females renest on the same territory, or when some females leave and others enter the territory within the season. Renesting in response to predation and other forms of nest failure is common in red-winged blackbirds, and there is some movement of females between territories within seasons (Picman 1981, Beletsky and Orians 1991). In our example, nests 5 and 6 may be renesting attempts by the females responsible for nests 1 and 3, so that a harem size of six overestimates the maximum number of simultaneous mates, which is four.

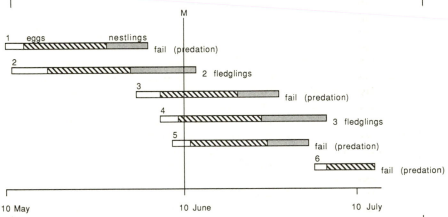

Figure 1.1 A hypothetical nesting chronology for one territory. Each nesting attempt is numbered (1–6), and the outcome of each attempt is shown. Open bars indicate nest construction, cross-hatched bars show egg laying and incubation, and stippled bars indicate the nestling period. Line "M" shows a date (10 June) on which the maximum number of active nests was achieved.

BOX 1.1 CONT.

METHOD 2

The total number of females nesting on a territory (e.g., Nero 1956a, Case and Hewitt 1963). This method requires that all females be marked so that they can be counted. The method will overestimate the maximum number of simultaneous mates if one or more early-settling females have left the territory by the time later ones enter it. In our example, suppose that each nest was started by a different female, but that females 1 and 2 had left t'e territory by the time female 6 entered; then Method 2 would give a harem size of six when the maximum number of simultaneous mates was five.

METHOD 3

The maximum number of simultaneously active nests on a territory (e.g., Weatherhead and Robertson 1977b, Yasukawa 1981a, Patterson 1991). This method can be used when nest chronologies are available but females are not individually marked. In practice, harem size is estimated by finding the vertical line (= date) that passes through the largest number of nest chronologies. In the current example, line M indicates a harem size of four. This method may underestimate harem size if some females remain on the territory when not nesting. In our example, the female responsible for nest 2 may have remained on the territory feeding fledglings and then later have built nest 6, in which case the maximum number of simultaneous mates would be five rather than four.

METHOD 4

The minimum number of females necessary to account for all nesting attempts on a territory (e.g., Holm 1973, Orians 1973, Searcy 1979a). This method is similar to Method 3 except that it assumes that a female remains on a territory without nesting for a certain period after either failure or success of a nesting attempt; for example, Searcy (1979a) assumed females remained for seven days after failure and for 20 days (including time spent feeding nestlings) after success. This method may overestimate harem size if the assumptions about how long females remain without renesting are wrong. In our example, if nest 5 were started less than seven days after nest 1 failed, Searcy (1979a) would estimate harem size as five; the maximum number of simultaneous mates might be four, either because a single female (renesting in less than seven days) was responsible for both nests 1 and 5, or because the female responsible for nest 1 left the territory before the female responsible for nest 5 settled.

The ideal way to estimate harem size (as we define it) would require marking females and constructing nesting chronologies, determining which females were associated with which nests, and observing directly when females were on territory before or after nesting attempts. As far as we know, no one has yet produced harem size estimates in this way. In the absence of this ideal, Methods 2, 3, and 4 seem to have comparable advantages.

BOX 1.1 CONT.

We applied all four methods to unpublished data gathered in 1992 from Yasukawa's upland population in southern Wisconsin; results are shown below.

Method	Mean Harem Size	Maximum Harem Size
1 (total # of nests)	2.2	10
2 (total # of females)	2.4	5
3 (# of simultaneously active nests)	2.2	5
4 (minimum # of females to account for nesting)	2.2	5

Methods 2, 3, and 4 obviously give similar results. Methods 3 and 4 give results that are particularly close; with these two methods, harem sizes were identical for 32 of 36 territories (89%) and differed only by one for the remaining four territories.

the norm rather than the exception; species with comparable (but not greater) levels of polygyny include fan-tailed warblers, yellow-headed blackbirds, and corn buntings (Table 1.2). Clearly, if we can speak in terms of varying degrees of polygyny, red-winged blackbirds are among the most polygynous of bird species.

So far we have been applying a social definition of polygyny to red-winged blackbirds; what about a genetic definition? Social and genetic criteria for mating systems would be completely congruent for redwings if all the females nesting on any territory copulated exclusively with the owner of that territory. To the extent that females perform copulations with males other than the territory owner ("extrapair copulations," or "EPCs"), and these copulations are successful in siring offspring, the social and genetic definitions are not congruent. Occasional EPCs were noted by early workers (Allen 1914, Beer and Tibbits 1950), but it was not until 1975 that hard data began to appear on the magnitude of the phenomenon. These first data came from an ingenious study in which Bray et al. (1975) sterilized territorial male redwings using vasectomy. The intent of the study was to test the effectiveness of male sterilization in population control. To the surprise of all, nearly half of the clutches on territories of sterilized males had at least one fertile egg. Sperm storage by females might provide a partial explanation for the fertility of these clutches, but Bray et al. minimized the importance of sperm storage by including only clutches started at least five days after vasectomy.

Table 1.1

Degree of Polygyny in Different Populations of Red-winged Blackbirds.

Method	Mean Harem Size	Maximum Harem Size	% Polygyny	N^d	Habitat	Locale	Reference
1	2.6	—	—	111	marsh	Illinois	Smith 1943
2	2.0[a]	3	80[a]	25	marsh	Wisconsin	Nero 1956a
2	3.5[b]	6	—	42	marsh	California	Orians 1961
2	2.2[a]	—	85[a]	42	marsh	New York	Case & Hewitt 1963
2	1.8[a]	2	—	26	upland	New York	Case & Hewitt 1963
4	2.9[a]/3.0[b]	6	83[a]/85[b]	104	marsh	Washington	Holm 1973
4	2.6[a]/3.5[b]	9	63[a]/83[b]	16	marsh	Costa Rica	Orians 1973
4	2.7[a]	4	80[a]	10	upland	Iowa	Blakley 1976
3	2.8[a]	9	—	97	marsh	Ontario	Weatherhead & Robertson 1977a
4	4.6[a]/5.0[b]	15	90[a]/97[b]	107	marsh	Washington	Searcy & Yasukawa 1983
3	4.1	6	87	39	marsh	Ontario	Weatherhead 1983
4	4.0[c]	—	—	64	marsh	California	Ritschel 1985
4	6.2[c]	—	—	13	upland	California	Ritschel 1985
2	2.7[a]/2.7[b]	5	77[a]/77[b]	22	marsh	Pennsylvania	Searcy 1988a
2	4.1[b]	14	86[b]	729	marsh	Washington	Orians & Beletsky 1989
3	3.2[b]	7	91[b]	34	marsh	Indiana	Patterson 1979
2	2.4	9	—	51	marsh	Alberta	Whittingham & Robertson 1994
3	1.7[b]	4	49[b]	43	mixed	New York	Westneat 1993a

NOTE: For methods of determining harem sizes, see Box 1.1.
[a]Based on all territorial males.
[b]Based on mated males only.
[c]Sample of males not specified in original reference.

Table 1.2

Degree of Polygyny Shown in Other Socially Polygynous Species of Birds.

Species	% Polygyny[c]	Mean Harem Size	Maximum Harem Size	N	Reference
indigo bunting	10[a]/11[b]	1.0[a]/1.1[b]	2	141	Carey & Nolan 1979
house wren	10[a]	1.1[a]	2	31	Johnson & Kermott 1991
Tengmalm's owl	10[a]/13[b]	0.9[a]/1.1[b]	3	374	Korpimäki 1991
pied flycatcher	11[a]/13[b]	1.0[a]/1.1[b]	2	79	Alatalo et al. 1990
winter wren	11[a]/13[b]	1.0[a]/1.1[b]	2	44	Wesolowski 1987
northern harrier	16[b]	1.3[b]	5	70	Simmons et al. 1986
wood warbler	17[a]/18[b]	1.1[a]/1.2[b]	3	137	Wesolowski 1987
prairie warbler	18[b]	1.2[b]	3	115	Nolan 1978
blackpoll warbler	20[a]/25[b]	1.0[a]/1.3[b]	2	35	Eliason 1986
marsh wren	25[a]/30[b]	1.1[a]/1.3[b]	2	80	Verner & Engelsen 1970
great reed warbler	27[a]/33[b]	1.1[a]/1.3[b]	2	37	Catchpole 1986
	46[a]/65[b]	1.2[a]/1.8[b]	4	37	Ezaki 1990
bobolink	29[a]/37[b]	1.1[a]/1.4[b]	3	139	Wittenberger 1978
Brewer's blackbird	31[b]	1.4[b]	4	114	Williams 1952
western meadowlark	33[a]/37[b]	1.2[a]/1.4[b]	2	33	Dickinson et al. 1987
dickcissel	35[a]/44[b]	1.2[a]/1.5[b]	3	31	Zimmerman 1966
fan-tailed warbler	60[a]/83[b]	2.7[a]/3.7[b]	11	111	Ueda 1984
yellow-headed blackbird	87[b]	2.8[b]	5	46	Lightbody & Weatherhead 1988
corn bunting	88[b]	2.7[b]	7	33	Ryves & Ryves 1934a, b

[a]Based on all territorial males.
[b]Based on mated males only.
[c]The percentage of males that mated polygynously.
[d]N is the number of males sampled.

The vasectomy data made it clear that substantial numbers of EPCs must occur but gave little indication of the relative frequency of extrapair versus intrapair copulation. This problem was addressed by two separate studies that directly observed copulations by marked individuals in the same population in central Washington (Monnett et al. 1984). Both studies found extrapair copulations to be less frequent than intrapair ones, amounting to 12% of the 58 copulations in one study (Monnett et al. 1984) and 16% of 56 copulations in the other (Emily Davies pers. comm.). A more recent study on a New York State population found an even lower incidence of EPCs, which constituted 6% of 71 total copulations (Westneat 1992a). The authors of all these studies made careful attempts to obtain an unbiased sample of copulations, but there is still a real possibility of bias in that extrapair copulations may be more rapid or more cryptic than intrapair ones. Also, it may be that the timing of EPCs is biased either toward or away from each female's period of maximum fertility. Thus data are also needed on which males are actually fertilizing eggs.

This need has been filled only very recently, through studies employing DNA fingerprinting. This is a technique from molecular biology, in which DNA (usually extracted from blood in birds) is cut into fragments and then spread on a gel. The fragments assort on the gel according to size, and one or more radioactive probes are used to label fragments containing particular highly variable sequences of nucleotides (Jeffreys et al. 1985, Burke and Bruford 1987). An audioradiograph is made of the gel, producing a series of bands, each of which is inherited by an offspring from its parents in a Mendelian fashion. If the band pattern ("fingerprint") is determined for a nestling and both of its parents, then all or almost all of the bands in the nestling should match bands in the parents. If instead the nestling has many "novel" bands, this indicates that one of the putative parents is actually not a genetic parent; which one can be determined by the degree of band sharing between the nestling and each parent. Because many bands are produced, and each is fairly rare in the population, this is a very powerful method of excluding individuals as parents. In fact, in cases where one parent is excluded, the true parent can often be identified from among local adults by its degree of band sharing with the nestling.

Gibbs et al. (1990) used DNA fingerprinting and related methods to examine genetic parentage for 36 redwing broods on three small marshes in southern Ontario. Seventeen of the 36 broods (47%) turned out to contain one or more young not sired by the territory owner. Generally the territory owner sired some of the young even in these 17 broods, whereas the rest were sired by an additional male. No instances were found in which three or more males sired young in one nest. Altogether, 31 of 111 young (28%) were products of extrapair copulation. Gibbs et al. were able to show that all or nearly all extrapair fertilizations were accomplished by other territorial males rather than by floaters. In all cases, the female attending the nest proved to be the genetic mother of all the young in the brood.

Westneat (1993a) also used fingerprinting to examine parentage of red-wing young during a two-year study of a population inhabiting a mixed marsh and upland site in New York. Twenty-eight of 68 broods (41%) proved to have at least one offspring sired by a male other than the territory owner. Of 235 young, all were found to be the genetic offspring of the female caring for the nest, whereas 55 (23%) were not the genetic offspring of the male holding the territory at the time of mating. Westneat (1993a) was able to show that at least 33 of the 55 extrapair fertilizations (60%) were accomplished by other territory owners. The two fingerprinting studies agree, then, in finding a considerable incidence of extrapair fertilization, in the range of 20 to 30%, and no evidence of intraspecific brood parasitism (i.e., of females laying eggs in nests of conspecific females).

It is clear that both sexes of red-winged blackbirds often share gametes with more than one member of the opposite sex within a breeding season. The great majority of territorial males have more than one female nesting on their territory with which they share gametes, and in addition these territorial males may share gametes with females nesting on other territories as well. About half of the females share gametes only with the owner of their nesting territory, whereas the other half share gametes with the territory owner plus one additional territorial male. Clearly, then, the genetic mating system is not simple polygyny but rather should be termed polygynandry. We conclude that red-winged blackbirds are socially polygynous and genetically polygynandrous.

The Evolution of Polygyny

Over the years, many hypotheses have been suggested to explain the occurrence of social polygyny in birds. Prior to the 1960s, the most popular hypothesis was that polygyny is produced by a female-biased sex ratio. Some early formulations of this "skewed sex ratio" hypothesis were not explicit about the mechanism leading from a female-biased sex ratio to polygyny (Skutch 1935), whereas others proposed mechanisms that today have little appeal. Thus Ryves and Ryves (1934b) stated that "corn-buntings definitely are ordained to be polygamous and . . . to this end, Nature arranges for the production of females in considerable excess of males." Only slightly more respectable by today's standards is the group selectionist mechanism proposed by L. Williams (1952), who suggested that polygyny in Brewer's blackbirds "is of advantage to the species in that when an excess of females occurs, the proportionately fewer breeding males may be able to accommodate these extra females." This hypothesis is group selectionist in that a trait, in this case polygyny, is proposed to have evolved because of its benefit to a population or species; most evolutionary biologists today would hold that, in general, traits evolve primarily in response to their costs and benefits to individuals rather than groups (G. C. Williams 1966). Although we may reject the logic of these early formulations, it is possible to suggest a mechanism for the skewed-sex-ratio hypothesis that is evolutionarily reasonable. If the sex ratio is female biased, all or substantially all of the males in a population may obtain a mate while some females remain unmated. These unmated females are then faced with the choice of either settling on the territory of an already-mated male or foregoing breeding altogether. Under such circumstances it may be individually advantageous for females to choose polygyny.

Dissatisfaction with the skewed-sex-ratio hypothesis arose as polygynous species were found with even sex ratios, and monogamous species with biased sex ratios (Orians 1961, Verner 1964). Alternative hypotheses then began to appear, most notably what later became known as the "polygyny threshold" model. This was first stated as a simple verbal hypothesis by Verner (1964), who proposed that "a female selecting a mate . . . might rear more young by pairing with a mated male on a superior territory than with a bachelor on an inferior one, notwithstanding the fact that she would obtain less help from her mate." Here Verner assumed that polygynous mating would be advantageous to males, whereas females mating with already-mated males would experience a cost of polygyny, principally because their young would receive less male parental care. Females would mate polygynously only when compensated for the cost of losing male help by obtaining a sufficiently superior nesting territory. Verner and Willson (1966) elaborated the logic of the hypothesis, while introducing the term "polygyny threshold":

> [P]olygyny can be advantageous for females if, within the limited area from which a female is likely to select a mate, the difference between two males' territories is sufficient that a female is able to rear more offspring on the better territory, by herself, than she could rear on the poorer one even with full assistance from the male. This difference between territories can be regarded as a "polygyny threshold," since it is likely that polygyny will be favored by natural selection whenever the difference exceeds a certain level.

This verbal model was subsequently cast in graphical form by Orians (1969a). The graphical model (Figure 1.2) assumes that female fitness, measured as seasonal reproductive success, is positively correlated with the quality of the environment in which the female settles. The model assumes a cost of polygyny, which is incorporated in the graph by drawing the expected fitness curve for the second female settling on a territory below the

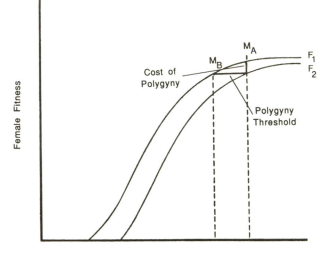

Figure 1.2 The polygyny threshold model (PTM) of Orians (1969a), with modifications following Wittenberger (1976). The curve marked "F_1" gives the expected fitness of the first female to settle on a territory as a function of Breeding Situation Quality (BSQ), a combination of the quality of the male and his territory. The "F_2" curve is the fitness function for the second female to settle on a territory. The cost of polygyny incurred by a second female on any territory is the difference between F_1 and F_2 for that territory (the figure illustrates the cost of polygyny for M_A's territory). M_A is the male with the best breeding situation overall. The first female to choose should mate with M_A, and subsequent females are forced to choose males with progressively lower BSQs. Females should continue to mate monogamously (i.e., they should follow the F_1 curve) until the unmated male with the highest available BSQ is M_B. At this point, the next female to choose will do as well by becoming the second female on M_A's territory as by becoming the first female on M_B's (i.e., the expected fitnesses for the two breeding situations are equal). The polygyny threshold is the minimum difference in BSQ that justifies a decision to mate with an already-mated male (i.e., that exactly compensates the second female for the cost of polygyny). Reprinted with permission from *American Naturalist*, copyright 1969 by the University of Chicago.

fitness curve for the first female. The fitness curve for a third female would be drawn below that of the second female, etc. Orians (1969a) expanded the list of factors producing a cost of polygyny to include increased competition for resources among the females sharing a male's territory, and the possibility that higher densities of females would attract more predators. Females are assumed to choose whatever territory gives them the highest expected fitness. Polyygny occurs when the difference in environmental quality between the best mated and the best unmated male is large enough to compensate females for the cost of polygyny. This difference in environmental quality that just justifies the decision to mate polygynously is the polygyny threshold.

The polygyny threshold model (PTM) has been extremely influential, and for substantial reasons. First, this was for many years the only hypothesis that provided an explanation of polygyny that was fully compatible with the emerging consensus on the predominance of individual selection. Second, the model in its graphical form remains to this day the single most fully worked out explanation of polygyny. Finally, the model has the virtue of elegant simplicity, which gives it great intuitive appeal. In fact, such was the appeal of the model that for many years alternative hypotheses were rarely considered.

Some of the subsequent theoretical work proposed extensions or elaborations of the PTM. Wittenberger (1976) suggested modifying the model to incorporate the possibility that qualities of the male as well as of his territory might affect female fitness and female choice. This modification is incorporated into the graphical model by labeling the X-axis "breeding situation quality," a variable that combines male and territory quality. The "sexy son" hypothesis of Weatherhead and Robertson (1979) argues the potential importance of male attractiveness ("sexiness") in determining breeding situation quality. Weatherhead and Robertson suggested that if sons inherit the sexiness of their fathers, then a female choosing a polygynous male might be compensated for the cost of polygyny in part by the enhanced attractiveness of her sons. Altmann et al. (1977) suggested another extension of the PTM, a "cooperative" version of the model in which female fitness increases with increasing harem size, at least over some range of harem sizes.

More recently, several alternatives to the PTM have been proposed. Some of these recent models apply to species, such as the pied flycatcher, in which individual males defend multiple territories (i.e., are "polyterritorial"). Here it is assumed that males benefit from polygyny, whereas females experience a cost of polygyny for which they are not compensated (Alatalo and Lundberg 1990). The "deception" hypothesis proposes that females in polyterritorial species choose already-mated males under these circumstances because they are unaware that the males are already mated (Alatalo et al. 1981). Such deception is plausible because the first mate of the already-mated male occupies a separate territory at some distance from the second (Alatalo and Lundberg 1990) and because the already-mated male behaves similarly to an unmated male when a female is examining his second territory (Searcy et al. 1991). An alternative, "search cost" hypothesis suggests that females in polyterritorial species are actually aware of male mating status but that they mate with already-mated males because the cost of searching for unmated males is greater than the cost of polygyny (Stenmark et al. 1988).

Another recent alternative to the PTM, proposed for monoterritorial species, is the "neutral mate choice" model (Lightbody and Weatherhead 1988). This model assumes that females pay no cost of polygyny and so settle randomly with respect to the settlement of other females. Female settlement is also assumed to be independent of features of the male and his territory. Female settlement, then, mimics a process in which each female is assigned a territory randomly. It should be intuitively obvious that as long as the sex ratio is not strongly male biased, such a process would produce an appreciable incidence of polygyny.

Davies (1989) has criticized polygyny models such as the PTM for assuming that mating systems can be explained by the choices of one sex only. Instead, he suggests that polygyny and other mating systems are the product of "conflicts of interest within and between the sexes." This viewpoint has arisen out of Davies's studies of polygyny in the dunnock, a species in which monogamy, polygyny, polyandry, and polygynandry may all occur in the same population (Davies 1992). Male dunnocks have highest success in polygynous associations, whereas females do best in polyandrous ones. Females as well as males are territorial, and polygyny occurs when a single male is able to defend an area containing the territories of two females. The "conflict of interests" approach seems to us to be most useful in species, like the dunnock, in which the spatial distribution of females is established first and the male distribution is superimposed. More typically, males establish their territories first, and here polygyny does seem likely to be explained by choices made by females rather than by conflicts of interest between the sexes.

Despite the gradual accumulation of diverse hypotheses to explain the occurrence of polygyny, much of the empirical work on polygyny in birds has continued to examine a single hypothesis at a time, usually the PTM. Examination of a single hypothesis leads to uncertainty in interpreting the results. Typically the data are partially, but only partially, in accord with the hypothesis, and it is not clear whether the partial fit should be regarded as confirming the model or whether the lack of a complete fit should cause the rejection of the model. Therefore, we have advocated a procedure of testing between alternative hypotheses, and to this end we have proposed a set of alternative hypotheses intended to be as complete as possible (Searcy and Yasukawa 1989). Included are both the hypotheses we have described above and some novel ones to fill in gaps. The set of hypotheses is classified hierarchically, in a way intended both to reflect the logic of the hypotheses and to aid in testing them (Table 1.3). As this is the set of hypotheses we will test with data on red-winged blackbirds (see chapter 6), we introduce the hypotheses in detail now.

All the hypotheses assume that polygyny is advantageous or at least neutral to males. This assumption seems reasonable, but one can imagine circumstances under which it might be invalid. We will discuss evidence bearing on this assumption for red-winged blackbirds in chapter 4. Assuming that males will accept polygynous mating, the question becomes why females mate polygynously (Orians 1969a). The first dichotomy in the hier-

Table 1.3
A Hierarchical Classification of Models of Polygyny.

I. *Male coercion.* Males force females to mate polygynously.
II. *Female choice.* Males cannot force females to mate polygynously.
  A. *No cost.* There is no cost to polgyny (female fitness does not decease with increasing harem size).
    1. *Benefit.* Female fitness increases due to increasing harem size.
      a. *Directed choice.* Females choose mates according to breeding situation quality.
      b. *Random choice.* Females choose mates randomly.
    2. *No benefit.* Female fitness does not change harem size.
      a. *Directed choice.*
      b. *Random choice.*
  B. *Cost.* Female fitness decreases due to increasing harem size.
    1. *Skewed sex ratio.* Females must pay the cost of polygyny because of a lack of unmated males.
    2. *Balanced sex ratio.* Unmated males are available when females choose mated ones.
      a. *Compensation.* Females are compensated for the cost of polygyny by acquiring a mate of high breeding situation quality.
      b. *No compensation.* Females are not compensated for the cost of polygyny.
        i. *Search cost.* Females accept already-mated males because searching for an unmated male is costly.
        ii. *Deception.* Mated males conceal their mating status from females.
        iii. *Maladapted females.* Females knowingly choose to mate with mated males even though more advantageous strategies are open to them.

SOURCE: Searcy and Yasukawa (1989).

archy of models is between the "male coercion" model, which assumes that males are able to force females to mate polygynously, and "female choice" models, which assume male coercion is not effective. This dichotomy is placed first not because male coercion is of proven importance (it is not, at least in birds) but because if male coercion is operative, the more elaborate assumptions made by the other models are unnecessary.

Within the female-choice category, the next dichotomy is between "no cost" and "cost" models. No-cost models assume females do not experience a cost of polygyny, whereas cost models assume they do. Within the no-cost category, there is a further division between "benefit" models, which assume that females benefit from the presence of other females on the males' territories, and "no benefit" models, which assume females experience no such benefit. Both the benefit and no-benefit models are further divided into "directed choice" and "random choice" categories. In directed-choice models, females choose males according to their breeding situation quality, whereas in random-choice models, females choose males without regard to male or territory characteristics. Our no-cost, no-benefit, random-choice model is exactly equivalent to the neutral-mate-choice model (Lightbody and Weatherhead 1988). Again, an appreciable incidence of polygyny is expected under such a random-choice model, whereas if one adds either a benefit of polygyny or directed choice, or both, the degree of polygyny should increase. The benefit-directed-choice model is somewhat similar to Altmann et al.'s (1977) cooperative version of the polygyny threshold model.

Within the cost models, the first division is between the "skewed sex-ratio" hypothesis and the "balanced sex-ratio" models. The skewed-sex-ratio hypothesis corresponds to the evolutionarily reasonable version of the hypothesis of that name discussed above, in which females choose to pay the cost of mating with an already-mated male because of a lack of unmated males. Balanced-sex-ratio models assume that unmated males are available; these models are divided into "compensation" and "no compensation" alternatives. In compensation models, females mating with already-mated males are compensated for the cost of polygyny by acquiring superior breeding situation quality. Thus, the PTM is a compensation model and can be considered synonymous with the class of all compensation models if it is interpreted to allow all types of fitness compensation.

The remaining hypotheses are ones in which females pay a cost of polygyny despite the availability of unmated males and despite the lack of compensation. The deception hypothesis (Alatalo et al. 1981) proposes that females make these poor choices because they are unaware that the chosen males are already mated. The search-cost hypothesis (Stenmark et al. 1988) proposes that females are aware of male mating status but that the cost of searching for an unmated male is greater than the cost of polygyny. The final hypothesis, the "maladapted female" model, proposes no rationale to explain the poor choices of females choosing already-mated males; these females are making mistakes that they have sufficient information to avoid. This last hypothesis has little intuitive appeal, perhaps, but it may well be a valid explanation of polygyny in some cases, especially those in which females choose mates under unnatural or unusual circumstances (Petit 1991, Johnson and Kermott 1991).

All the hypotheses discussed so far take a rather short-term view of polygyny; they consider whether females should or should not mate with already-mated males, given existing conditions in terms of the cost of polygyny, the extent of female cooperation, sex ratios, search costs, etc. Thus, most of the factors that might influence mating decisions are assumed to be fixed. This assumption is realistic over the short term, but over the long term all these factors can change, and the changes may both affect and be affected by the mating system. For example, as a species becomes more polygynous, the relative payoff of mating effort compared to parental care may increase for males, with males consequently evolving to lessen further their contribution to parental care. Decreases in average male parental care should lessen the cost of polygyny because secondary females foregoing male parental help would be sacrificing a less important resource. Decreases in the cost of polygyny would lead to further increases in the degree of polygyny, etc. A second example is that as a species becomes more polygynous, females may be selected to become more cooperative, to lessen the cost of aggressive interactions among females forced to coexist. This change would again lower the cost of polygyny and would accelerate the species' evolution toward polygyny.

In this book we will pay most attention to the short-term models of polygyny, primarily for a practical reason: we know, or can find out about, present-day conditions, but we can only make educated guesses about the conditions that prevailed when a species began to evolve toward polygyny. Nevertheless, the longer-term questions are intriguing, and we will return to them later.

Sexual Selection

As one of our themes is the effects of sexual selection on red-winged blackbirds, we need to introduce certain concepts in sexual selection theory. We will start by defining sexual selection itself. Darwin's own definitions of sexual selection were never entirely satisfactory, in that they never completely meshed with his use of the term. Thus in *The Origin of Species*, Darwin (1859) stated that sexual selection "depends, not on a struggle for existence, but on a struggle between the males for possession of the females; the result is not death to the unsuccessful competitor, but few or no offspring." The problem with this definition is that it seems to restrict sexual selection to male-male aggressive competition, whereas it is clear from Darwin's (1859, 1871) examples that he wished also to include mate choice by both females and males. In *The Descent of Man and Selection in Relation to Sex* (vol. 1, p. 256), Darwin (1871) veered in the other direction, defining sexual selection too broadly, as "the advantage which certain individuals have over others of the same sex and species, in exclusive relation to reproduction." This definition seems to include many types of reproductive advantage, which Darwin then goes on to exclude explicitly, such as advantage due to better parental care or to modifications of the primary sex organs. A modern definition of sexual selection is selection due to variation in mating success. Although this definition differs from those stated by Darwin, it closely approximates the way in which he actually used the term.

A much-debated point in sexual selection theory has been the relationship that should be recognized between sexual selection and natural selection. Some authors have argued forcibly that we should consider natural and sexual selection to be separate (Arnold 1983, Ghiselin 1974), as Darwin himself clearly did. Darwin (1859) thought of natural selection as selection for success in the "struggle for life," which would increase fitness in the sense of improving designs for survival. Sexual selection, which favors traits enhancing mating success, could then be considered separate from natural selection. From a present-day viewpoint, the difficulty with this formulation is that most biologists have discarded the design-for-survival view of fitness in favor of the population geneticists' concept of fitness as success in passing genes to future generations. If one accepts this genetic definition of fitness, and if one defines natural selection as selection due to variation in fitness, then sexual selection becomes a subset of natural selection: improving mating success is just another way of enhancing the passage of genes. Relegating sexual selection to a position as a subset of natural selection, however, is one of the heresies against which Darwin's adherents have railed most strongly.

There is usually no right or wrong in such semantic arguments, so we will simply state the conventions we adopt in this book. We will use fitness in the population geneticists' sense, as success in passing genes. Sexual selection will mean selection due to variance in mating success, and we will consider sexual selection to be distinct from natural selection. This forces us to adopt a narrow (and awkward) definition of natural selection as selection due to variance in all components of fitness other than mating success (Arnold 1983).

It is generally agreed that in most species, sexual selection acts more strongly on males than on females, and that considering males only, sexual selection acts more strongly in polygynous species than in monogamous. The logic is that the opportunity for sexual selection ought to be proportional to variance in mating success and that variance in mating success is generally greater among males than among females, and greater among males in polygynous species than in monogamous ones. The argument has its weak points. For one thing, there is substantial disagreement on whether variance in mating success is an adequate measure of the opportunity for sexual selection, and if not, about what alternative measures should be used (Arnold and Wade 1984, Clutton-Brock 1988). For another, the effect of sexual selection depends not only on the opportunity for sexual selection but also on the strengths of correlations between phenotypic traits and mating success and on the heritability of the phenotypic traits (Clutton-Brock 1988). We will return to these issues later. Nevertheless, for now we will accept the proposition that sexual selection has acted more strongly on male red-winged blackbirds than on either conspecific females or males of monogamous species.

Mirroring the division between natural and sexual selection is another, traditional dichotomy, that between intrasexual and intersexual selection. Intrasexual selection refers to competition, usually aggressive, between individuals of one sex for mating opportunities. When applied to sexual selection acting on males, intrasexual selection is termed "male-male competition." Intersexual selection refers to mate choice, i.e., preferences shown by members of one sex for particular individuals of the opposite sex. Intersexual selection acting on males is termed "female choice of mates." In practice it is often impossible to subdivide the effects of sexual selection into intersexual and intrasexual components. One often cannot make a clean separation between variance in mating success due to within-sex aggressive competition and variance due to mate choice. Instead, sexual selection commonly acts in a linked, two-stage process: males compete aggressively with each other, and females choose the winners of the competition. Here the two components of sexual selection each rely on the other. Female choice is dependent on the outcome of male-male competition, and the goals of male-male competition are determined by female preferences. Even though the two components of sexual selection are not cleanly separable in this situation, the distinction between intrasexual and intersexual selection remains extremely useful.

A pattern in which males have better developed displays, ornaments, and weaponry than females is quite common among animals in general and is clearly seen in red-winged blackbirds. It was to explain the evolution of this pattern that Darwin formulated the theory of sexual selection. Darwin (1871) acknowledged that the occurrence of some sexually dimorphic traits could be explained by ordinary natural selection, as for example, dimorphism in those traits "connected with different habits of life" (what we today might call ecological adaptations) and those directly connected with the production and transferal of gametes. Most dimorphic traits, however, could not be explained in this way. As Darwin (1871, vol. 1, p. 257) wrote (in specific reference to dimorphic organs of sense and locomotion):

> [I]n the vast majority of such cases, [these traits] serve only to give one male an advantage over another, for the less well-endowed males, if time were allowed them, would succeed in pairing with the females; and they would in all other respects, judging from the structure of the female, be equally well adapted for their ordinary habits of life. In such cases sexual selection must have come into action, for the males have acquired their present structure, not from being better fitted to survive in the struggle for existence, but from having gained an advantage over other males, and from having transmitted this advantage to the male offspring alone. It was the importance of this distinction which lead me to designate this form of selection as sexual selection.

A few lines later he added:

> There are many other structures and instincts which must have been developed through sexual selection—such as the weapons of offence and means of defense possessed by the males for fighting with and driving away their rivals—their courage and pugnacity—their ornaments of many kinds—their organs for producing vocal or instrumental

music. . . . That these characters are the result of sexual and not of ordinary natural selection is clear, as unarmed, unornamented, or unattractive males would succeed equally well in the battle for life and in leaving a numerous progeny, if better endowed males were not present.

Darwin, then, proposes that dimorphism in traits associated with display, ornamentation, and fighting have evolved due to sexual selection, in most cases acting on males. One of our goals in this book is to test this general theory with respect to red-winged blackbirds.

Grafen (1988) has distinguished two approaches to studying the effects of selection in general. One approach is to investigate the current action of selection, what Grafen (1988) calls "selection in progress." With respect to sexual selection, this approach requires relating natural variation in the traits of interest to variation in mating success. It is also necessary to show that the measured variation in mating success leads to variation in reproductive success, preferably lifetime reproductive success (see Endler [1986] for a cogent discussion of how such demonstrations should be constructed). The contrasting approach is to study "adaptations," traits that serve some function for the organism and that are the products of past selection. Adaptations may not be subject to current selection, at least not directional selection, so the methods used to study selection in progress may tell us little about adaptation. Other methods may be more appropriate, particularly ones in which the traits in question are manipulated and rather proximate effects are measured. In terms of sexual selection, these proximate effects would usually be effects on male-male competition and female choice. This kind of experiment gets at the question of why the trait is adaptive.

Endler (1986) and Grafen (1988) critique the methods used in studying selection in progress. For the most part, the criticisms have to do with the general problem of inferring causality from correlations. As one example, it is possible for a favorable environment to cause many phenotypic traits to change in males, while also directly affecting mating success. Then correlations would be found between the male traits and mating success, even if the male traits themselves had no effect on mating success. Beyond these criticisms of methods, Grafen (1988) suggests that most evolutionary biologists are more interested in adaptation anyway, that we would rather know for what function a trait has evolved than whether selection is currently acting on it. We agree with Grafen on this last point, but nevertheless we will consider sexual selection in progress, as well as whether and how traits are adaptive under sexual selection.

Prospectus

Throughout this book we will attempt to summarize what is known about polygyny and sexual selection in red-winged blackbirds, but we will also point out what is poorly understood, pose some questions yet to be answered, and suggest some potentially valuable new studies. We will cite many of the published studies on red-winged blackbirds and will also include some of our own unpublished data gathered in five different study areas: (1) Turnbull National Wildlife Refuge, Spokane County, in eastern Washington State (see Searcy 1979a); (2) marshes and roosts around Millbrook, Dutchess County, in southern New York State (Searcy 1981), especially in the Cary Arboretum (Yasukawa 1981c); (3) marshes near the Pymatuning Laboratory of Ecology, Crawford County, in northwestern Pennsylvania (Searcy 1988a); (4) Yellowwood State Forest, Brown County, in southcentral Indiana (Yasukawa 1981b); and (5) Newark Road Prairie, Rock County, in southcentral Wisconsin (Yasukawa et al. 1987a).

The book is essentially divided into two sections, the first of which deals with the explanation of polygyny in red-winged blackbirds. This section starts with a chapter (chapter 2) on parental care, a topic important to models of polygyny in that the relative contribution of the sexes to parental care in large part determines the costs and benefits to females of sharing mates. We next turn to a discussion of territoriality in males (chapter 3), because the system of male territories underlies the polygynous mating system, and the characteristics of these territories may have a large effect on females in deciding whether to mate polygynously. Chapter 4 deals with female reproductive success, in particular concentrating on the aspects of males and their territories that affect female fitness. Chapter 5 considers the evidence on which of these aspects of males and territories actually influence female settlement, information that is crucial in testing many of the polygyny hypotheses. The first section concludes with chapter 6, which integrates much of the information in chapters 2 through 5 in testing the models of social polygyny set out in this chapter.

The second half of the book is devoted to the consequences of polygyny. Chapter 7 discusses sexual selection in progress; here we attempt to measure the current strength of sexual selection and to discover what male traits it favors. Chapter 8 considers whether traits that are sexually dimorphic in red-winged blackbirds, specifically visual and vocal traits, size, and aggressiveness, have evolved in males as adaptations to sexual selection. In chapter 9 we discuss the effects of polygyny on females, largely in terms of adaptations to living together in harems. Chapter 10 concludes the book with a summary of our conclusions on polygyny and sexual selection in red-winged blackbirds, a discussion of how well these conclusions generalize to other species, and some suggestions for future research.

Even though there are about 1000 theses and published papers on red-winged blackbirds, there are still many unanswered questions that we find intriguing. We hope some of our readers will agree and will be stimulated to study red-winged blackbirds themselves.

2 Parental Care

Parental care and mating system have a reciprocal relationship, each influencing the other. On the one hand, the form of the mating system affects the balance of costs and benefits of performing parental care. A major cost of parental care to males is lost opportunity to perform other activities, such as mate attraction, mate guarding, extrapair courtship, and territory defense. In general, the fitness payoffs of these alternative activities increase as the degree of polygyny increases. By influencing the benefit of these alternative activities, the degree of polygyny of the species can affect the amount of effort that males devote to parental care.

On the other hand, the relative contributions of the sexes to parental care can influence the form of the mating system. As we have seen, many hypotheses for the evolution of polygyny assume that females pay a cost of polygyny, i.e., a female's fitness is reduced because her mate already has other females. Perhaps the most important source of such a cost is sharing of the male's parental care. The more females that are mated to a given male, the less help each will obtain in raising her young. The amount of parental care contributed by the average male influences the magnitude of this cost of polygyny, which in turn can determine how polygynous the species is.

The cost of sharing male parental care depends not only on the magnitude of the male effort but also on the form. Some forms of parental care are thought to be "shareable," in the sense that care given to one offspring does not reduce the parent's ability to invest in other offspring. Other types of parental care are "nonshareable," meaning that care given to one offspring reduces the care given to others (Perrone 1975, Wittenberger 1981). The temporal limits of nonshareable care are somewhat open to debate. For example, it is clear that an item of food given to one nestling cannot be given to a second, so that feeding young in one brood reduces the amount of food received by the young in a second, concurrent brood. It is also possible, however, that young in later broods will not be fed by the male because his prior feeding effort was too stressful. In any case, it is obvious that a distinction between shareable and nonshareable parental care is important for the evolution of mating systems because nonshareable parental care produces a cost of polygyny, whereas shareable parental care does not.

It is also possible for male parental care patterns to influence the frequency of extrapair fertilizations, and vice versa. If males can assess their paternity in each brood, they may be selected to provide care preferentially to those broods in which a high proportion of the young are their own genetic offspring. Furthermore, if males allocate parental care according to paternity, this would provide a major selective cost to females of performing extrapair copulations and might therefore affect the overall frequency of extrapair fertilization.

If the mating system influences the pattern of parental care, and the pattern of parental care influences the mating system, we have a chicken-and-egg problem: which came first? This question may not have a simple answer, in that the two may have coevolved, with small changes in parental care producing small changes in the mating system, and vice versa. Nevertheless, it is clear that to understand the present mating system, we need to understand the present system of parental care. In this chapter we describe the pattern of parental care in red-winged blackbirds. First, we discuss the parental care activities undertaken by females and males. Second, we examine factors that affect the magnitude of the male's contribution to parental care and how the male apportions his care among the members of his harem.

Female and Male Parental Care

Certain aspects of parental care are performed only by the females in red-winged blackbirds. First, females alone build the nests. The nest is bowl-shaped and is built of marsh vegetation or grasses, which are carefully woven around sturdy supporting plants. It is a fairly elaborate construction, with an outer basket fashioned usually of reeds or coarse grasses, caulked with material such as mud and rotten wood, and then lined with fine grasses. One nest, on dissection, contained 142 cattail leaves and 34 strips of bark in the outer basket and 705 pieces of grass as lining (H. B. Wood in Bent 1958). Obviously, it must take considerable time to gather this much material and weave it together, and in fact females typically spend four to six days building a nest (Allen 1914, Beer and Tibbitts 1950). In some cases, however, females can complete a nest in less than two days, although such nests lack the mud caulking and grass lining and seem flimsy when compared with more complete nests. Male red-winged blackbirds sometimes perform "symbolic" nest building, in which they manipulate nest material or push and prod at a nest, while being watched by a female, but males never make any real contribution to construction (Nero 1956a). Nest construction by the female only is typical of icterines in general, though there are exceptions where both sexes contribute, and even one species (the yellow-hooded blackbird) in which males construct nests by themselves (Orians 1985).

Second, females alone incubate. The usual incubation period is 11 days, although 10 and 12 days are also observed (Case and Hewitt 1963). In ex-

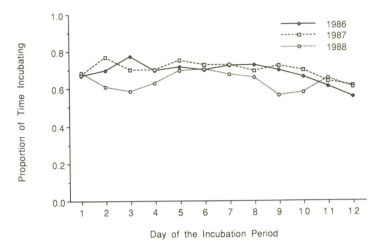

Figure 2.1 Proportion of time spent incubating by female red-winged blackbirds at Newark Road Prairie in Wisconsin, from Yasukawa (unpublished). Proportions shown are means calculated for 34, 38, and 31 females in 1986, 1987, and 1988, respectively.

ceptional cases, such as when incubating infertile eggs (or when given plaster eggs by curious ornithologists), females will persist for almost twice the normal period before abandoning the clutch. Incubating females are able to leave the nest at intervals to feed, but as shown in Figure 2.1, about 70% of their time is spent in the nest. Prior to incubation, female red-winged blackbirds develop brood patches to facilitate heat transfer to the eggs, and they undoubtedly expend energy, as well as time, on incubation. There is also some risk associated with incubation, as predators are known to take incubating females from their nests. According to Orians (1985), incubation by females only is universal among icterines, although in many other groups of birds, male incubation is common.

Third, females alone brood the young. The time investment in brooding is typically high early in the nestling period and then declines as the young develop the ability to thermoregulate on their own. Figure 2.2 illustrates this pattern for females on an upland site in Wisconsin for each of three years. Females at this site spent on average over 50% of their time brooding on the first day following hatching, but time brooding declined to less than 10% by day 7 or 8. During most of the nestling period, brooding would not place a major drain on either time or energy for females, but as with incubation, brooding may entail an increased risk of predation.

The remaining major aspects of parental care are guarding the eggs and young and feeding the young. Adults of both sexes participate in these activities but to varying degrees. We shall consider these aspects of parental care at length in the following sections.

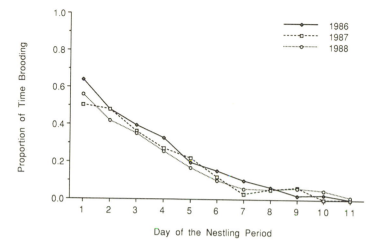

Figure 2.2 Proportion of time spent brooding by female red-winged blackbirds at Newark Road Prairie in Wisconsin, from Yasukawa (unpublished). Proportions shown are means calculated for 32, 30, and 37 females in 1986, 1987, and 1988, respectively.

Guarding Eggs and Young

Protecting eggs and young from predators can take two forms: the guarding individual can warn of the approach of the predator ("alerting") and/or it can attack the predator ("active defense"). Before either warning or defense can be accomplished, the guarding individual must discover the approaching predator, through "vigilance."

ANTIPREDATOR VIGILANCE

Male red-winged blackbirds with active nests on their territories spend considerable amounts of time engaged in what appears to be antipredator vigilance. Vigilant males sit on high, exposed perches and continually scan the sky. Of course, such vigilance undoubtedly has benefits other than nest guarding: males may be watching for intruding males, for newly arriving females, and/or for predators that pose a risk to themselves. Recent experiments by Yasukawa et al. (1992b) and Yasukawa (unpublished), however, indicate that vigilance is, at least in part, directed at guarding eggs and young.

In one set of experiments conducted at Newark Road Prairie in Wisconsin (Yasukawa unpublished), poles were placed near nests located in portions of the prairie that lacked natural, tall perches. Each pole provided a potential perch two meters above and five meters from a nest. In response to the addition of these perches, males significantly increased the amount of time spent perched within 10 meters of their nests, from 3.0% to 13.7% of the observation periods. When poles were placed in a 20-meter grid pattern, males used them in a nonrandom fashion, preferring to perch significantly closer to active nests (27 meters), on average, than would be expected (42 meters) if they chose perches at random with respect to nest locations.

In a second experiment conducted at Newark Road Prairie, a model crow was presented at redwing nests for five minutes and then removed. Males observed during the 30 minutes after such a presentation spent, on average, 36% of their time "in attendance" (perched within 10 meters) at the nest. By contrast, males spent only 19% of their time attending the same nests during randomly chosen control periods (Yasukawa et al. 1992b). Thus, once a nest was "discovered" by a predator, males significantly increased their vigilance. Males also increased their vigilance in response to the relatively mild threat presented by a human observer, spending significantly more time in attendance when a human observer was visible 50 meters from the nest (19%) than when the observer was hidden in a blind at the same location (7%) (Yasukawa unpublished).

There is also evidence that nests built close to tall perches are safer from nest predation than are nests with no nearby perches (see chapter 4) and that females prefer to place their nests near such perches (see chapter 5). These results, together with those described above, suggest that male vigilance is associated with nest guarding and that females seek out and benefit from nesting near perches that facilitate male vigilance.

Unlike males, female red-winged blackbirds cannot afford to spend considerable amounts of time scanning for predators from exposed perches once they begin to incubate (see Figure 2.1), and they do not do so. Nevertheless, females when sitting on the nest or moving about the territory sometimes spot an approaching predator before the male and are then the first to give warning or to attack.

ALERTING

Nesting red-winged blackbirds become agitated when humans or other potential predators approach active nests. One response that is typical of such agitation is calling by males and females. In habitats where nests are difficult to locate, researchers can use the intensity of calling as an aid to finding the nests, as one uses the cues "warmer" and "colder" to find a hidden object in a children's game. Knight and Temple (1988) reported that adult red-winged blackbirds gave seven different calls in response to experimental presentations of potential nest predators. In these experiments, females in general gave more alarm calls than males, about twice as many overall. The most common call in both sexes was the "check," which made up 60% of the calls given by males and 85% of those given by females. Two calls were

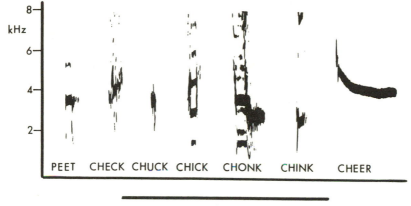

Figure 2.3 Spectrograms of the seven "alert calls" used by male red-winged blackbirds in a Washington marsh, from Beletsky et al. (1986). Specific call types are not used to refer to specific dangers, with the partial exception of the "cheer," which is sometimes termed the "hawk alarm call." Instead, males tend to call at a low rate during all activities and to match call type with neighbors. The primary signal of danger or disturbance is a change in the call type being used (see Beletsky et al. 1986).

produced by females only, "chatters" (1%) and "screams" (14%). Four calls were produced by males only, "growls" (3%), "peets" (1%), "seets" (4%), and "cheers" (33%).

Beletsky et al. (1986), using a somewhat different system for classifying calls, found that male red-winged blackbirds gave a total of seven call types: "peet," "check," "chuck," "chick," "chonk," "chink," and "cheer" (see Figure 2.3). All these call types are given in a variety of contexts, including not only the approach of predators but also situations such as courtship, male-male aggression, and feeding (Beletsky et al. 1986). Of the various calls, only the "cheer" is thought to refer to a specific class of predators, namely hawks; it is sometimes termed the "hawk alarm call" (Orians and Christman 1968). Knight and Temple (1988) found that the "cheer" was given in response to a taxidermic mount of a raccoon and a model of an American crow as well as to a mount of a red-tailed hawk. In our experience, "cheer" is also given in response to ornithologists checking nests. Thus "cheer" is not completely specific to hawks, but as Orians and Christman (1968) note, it is a variable call, and the highest pitched, thinnest versions do seem to be given only for hawks, in particular for accipiters, which pose the greatest danger to adult red-winged blackbirds of any predator. We have found as birdwatchers that listening for high-pitched hawk alarm calls is of great assistance when trying to add an accipiter to one's day list; redwings are better at spotting these predators than we are, but then they have a stronger motivation.

If the call types, for the most part, do not refer to specific classes of predators, why do males give more than one type? Beletsky et al. (1986) suggested that redwings have evolved a system in which switching between call types is used as an "alerting" system. As evidence, they showed that males increased the rate of switching between call types when exposed to a stuffed hawk. Further, Beletsky et al. (1986) demonstrated through observation and playback experiments that nearby males tended to match each other in call type and that when one male switched, other males tended to switch also. Thus, male redwings maintain an alerting system in which each territory owner gives repetitive calls at low rates in the absence of any disturbance, a male signals alarm by switching call types (and increasing call rate), and the alarm is spread by neighboring males switching also. Beletsky (1990) has shown that females attend to male call switching, increasing their own vocalization rates and, in some circumstances, remaining on the territory in response to playbacks of male calls with rapid switching rates.

One function of the alarm calls is to elicit active nest defense. Knight and Temple (1988) played four of the calls, female "check," female "scream," male "cheer," and male "growl," from a tape recorder placed on a territory but away from any active nest. "Checks," "screams," and "growls" all increased the likelihood that redwings would hover over the tape recorder, and in addition, "screams" attracted significantly more redwings to within five meters of the tape recorder. Alarm calls can also function to warn the young of the threat of predation. Knight and Temple (1988) found that "screams" caused nestlings to stop begging and crouch down in the nest, which presumably makes them more difficult to locate; the other alarm calls tested did not have this effect.

Active Defense

Red-winged blackbirds are well known for the aggressiveness of their active nest defense, as shown against humans. A newspaper once proclaimed "Yes! It's a Bird and It's Angry" in describing "dive-bombing" attacks by red-winged blackbirds nesting in a park. Any person who strays too near a redwing territory during the breeding season is likely to be attacked; naturally, researchers checking nest contents are particular targets. Anyone who has worked on breeding redwings must be familiar with the sensation of being struck on the back of the head by a male who has just performed a power dive from a height of 20 meters. Such attacks, though unlikely to cause even minor injury, are definitely disconcerting enough to exercise a certain deterrent effect. A wide range of other animals are attacked in similar manner: birds such as hawks, crows, jays, magpies, and herons, reptiles such as snakes, and mammals such as raccoons, minks, dogs, and even cows and horses. On occasion, redwings will go so far as to direct their attacks against cars and tractors.

Males in general play a stronger role in active defense than do females, although the strength of defense is extraordinarily variable among males. In their experiments with mounts and models of predators, Knight and Temple (1988) found that males on average approached closer to the predators than did females and were more likely than females to hover over the predator. Most importantly, males dove at and struck the predators much more often than did females, about nine times more often altogether. During natural acts of nest predation by minks, male redwings again approached more closely than females, although neither sex struck the predator, and defense seemed ineffective (Knight et al. 1985). In our experience, male redwings are far more likely to strike human observers than are females.

Feeding Young

Bringing food to offspring is characteristic of species of birds with altricial young. The great majority of altricial species are monogamous and show fairly equivalent effort by the two parents in feeding nestlings. The situation is more variable in polygynous species: in some there is a fairly equivalent effort by males and females, but in many polygynous species males undertake less provisioning than do females, and in others males make no contribution at all. One reason that red-winged blackbirds are interesting to students of parental care is that they show much of this variability within a single species.

Female red-winged blackbirds in all populations studied thus far provide food for their offspring. Among males, some individuals provision and some do not, and the frequency of provisioning varies greatly among populations. Anecdotal reports have long suggested that the frequency of male provisioning is higher in eastern than western populations. For example, Bent (1958) cited observations of an eastern redwing male feeding nestlings, whereas Orians (1961) and Payne (1969) reported that feeding by males in California was rare or nonexistent. More recent and detailed studies of male parental care tend to support a difference between eastern and western populations. Table 2.1 summarizes the quantitative data on the frequency of male feeding. In seven studies of eastern and midwestern populations, between 29 and 88% of the males fed nestlings. In contrast, one quantitative study of a far western population, in Washington, showed that only 6% of the males fed nestlings (Beletsky and Orians 1990). Interestingly, a population in Alberta, just east of the Rocky Mountains and thus intermediate along the east-west axis, had an intermediate level of male feeding (33%) (Whittingham and Robertson 1994). Further study of this east-west pattern might be very rewarding, especially an investigation of the transition zone that must exist in the mountains of the West.

Table 2.1
Frequency of Male Provisioning of Nestlings in Various Populations of Red-winged Blackbirds.

Locale	Habitat	Longitude	Percent of Males Provisioning	N	Reference
New York	mixed	74°	88	17	Yasukawa & Searcy 1982
New York	mixed	77°	66	75	David F. Westneat pers. comm.
Ontario	marsh	77°	80	50	Muldal et al. 1986
Ontario	marsh	77°	48	23	Eckert & Weatherhead 1987b
Ontario	marsh	77°	60	25	Muma & Weatherhead 1989
Ontario	marsh	77°	29	52	Weatherhead 1990a
Michigan	marsh	84°	38	34	Whittingham 1989
Indiana	marsh	87°	59	34	Patterson 1991
Wisconsin	upland	89°	87	79	Yasukawa et al. 1990
Alberta	marsh	113°	33	51	Whittingham & Robertson 1994
Washington	marsh	119°	6	472	Beletsky & Orians 1990
New York	mixed	77°	66	75	David F. Westneat pers. comm.

The frequency of male provisioning also may vary between sites in the same general area, and between years at the same site. In Alberta, 18% of males fed nestlings on one marsh and 77% on another only a few kilometers distant in the same year (Whittingham and Robertson 1994). In a single marsh in Michigan, Whittingham (1989) found that 60% of males fed nestlings in one year, whereas none did the following year. Yasukawa (unpublished) found that the percentage of males provisioning on a Wisconsin upland site varied from 32 to 94% over a 10-year period.

One factor that helps predict which males feed and which do not is male age, with older males, in general, being more likely to feed young. For example, in Beletsky and Orians's (1990) Washington population, the 29 males that fed nestlings had an average age of 4.3 years, compared to an average age of 3.6 years for territory owners in general. Patterson (1991) found the same trend in an Indiana population. This trend does not, however, explain why males in eastern and midwestern populations are more likely to feed than males in western populations, as there is no reason to think that the average age of male territory owners differs between the two regions.

A partial explanation for the regional difference in the frequency of male provisioning may lie in the somewhat larger mean harem sizes found in western populations (chapter 1), which may produce stronger selection favoring mate attraction and mate guarding at the cost of male parental care. Whittingham (1994) removed females from nests in Alberta and observed whether males took over the task of provisioning the nestlings. Of 11 males that had females in their fertile periods on their territories, none fed the young, whereas all 11 males that had no fertile females fed the young. On control territories (no female removed), none of 11 males fed nestlings when a fertile female was present, whereas 11 of 26 fed young when there was no fertile female present. The availability of unmated females (ones that had not yet settled) did not influence male provisioning. This study shows that male provisioning is strongly influenced by a tradeoff between provisioning on the one hand and mate guarding and courtship on the other; this is compatible with the idea that provisioning is less frequent in the West in part due to harems being larger there, because with larger harems males are more likely to have fertile females present at any given time.

Differences in food availability may also help explain regional differences in male provisioning. This idea is supported by Whittingham and Robertson's (1994) demonstration that males on a marsh surrounded by agricultural fields were more likely to feed nestlings than males in a marsh surrounded by woodland. Although the agricultural marsh produced more emergent insects than the woodland marsh, most food for nestlings was gathered off the marshes in surrounding areas, and the woodland was more productive than the agricultural fields, so overall food availability was lower on the agricultural marsh than on the woodland marsh (Whittingham and Robertson 1994). Females feeding alone in the woodland marsh could provide as much food for nestlings as two parents working together in the agricultural area. The impact of male feeding was greater in the agricultural marsh than in the woodland; females with male assistance raised 1.2 extra fledglings in the agricultural marsh compared to 0.6 extra fledglings in the woodland marsh. Thus, males may be more likely to feed where food is more limited and the benefit to their reproductive success is greater.

Males, even when they do provision, typically do so at only one nest on any day. As there are often several nests simultaneously active on the territory, many nests are not aided at all. Males usually do not begin provisioning until the young are three or four days old, whereas females of course begin feeding soon after hatching. Moreover, even at nests with older nestlings at which males are assisting, the female still typically brings more food than does the male. For example, Muldal et al. (1986) found that at male-assisted nests observed eight days after hatching in an Ontario population, 66% of the feeding visits were made by the female, versus only 34% by the male.

The overall conclusion, then, is that even in populations where male provisioning is common, females perform the bulk of feeding. Thus in Yasukawa et al.'s (1990) Wisconsin population, in which 87% of males fed nestlings, more than 80% of all feeding visits at nests of primary females were made by females, more than 90% of visits were by females at secondary nests, and virtually 100% were by females at other nests. Hurd et al. (1991) videotaped visits by parents at 18 redwing nests in Ontario and found that males contributed only 5% of feeding visits to nestlings. Teather (1992) videotaped at 10 nests, again in Ontario, and found only 3% of visits were made by males.

Two qualifications need to be added to these conclusions about the relative importance of male and female provisioning. One is that studies of provisioning have consistently used number of feeding visits to measure parental contributions, rather than measuring the actual amount of food delivered. Whittingham and Robertson (1994) found at their Alberta study sites that prey loads brought to nests by males were on average 27% larger (in mass) than those brought by females. If this holds true at other sites, then the relative contribution of males would be somewhat greater than suggested by rates of feeding visits. The second qualification is that studies of provisioning have concentrated on feeding of nestlings, while paying little attention to feeding of fledglings. Both male and female parents are known to feed the young after they leave the nest. Nero (1984) observed such provisioning continuing more than three weeks after the young had fledged, and after both the parent and the young had abandoned the territory. Quantitative data on feeding of fledglings is scanty. Beletsky and Orians (1990) found in their Washington population that males fed fledglings in 4.9% (23 of 472) of the observed male-years, making provisioning of fledglings just slightly less frequent than provisioning of nestlings (Table 2.1), but they warn that their estimate for fledglings is likely to be an underestimate. Searcy (unpublished) found that 74% of males (14 of 19) in a Pennsylvania population fed fledglings, which is within the range found for provisioning of nestlings in eastern North America (Table 2.1). Presumably, 100% of fledged broods are fed by females in both western and eastern populations. Thus, it may be that provisioning of fledglings follows the same pattern as provisioning of nestlings, with males providing less provisioning than females, and male provisioning being more common in eastern than western populations, but much more work is needed on this subject.

Apportionment of Male Parental Care

In some polygynous species, the females mated to a particular male nest asynchronously enough that there is little overlap in their respective nesting periods, and the male is able to help each female fully (Leonard 1990). In red-winged blackbirds, however, temporal overlap between females within harems is typically quite substantial. For example, Orians (1980) found that the mean interval between first and second clutches within territories was six days, and between second and third clutches approximately five days. Male redwings, then, are often faced with the problem of apportioning their parental care among simultaneously active nests.

In certain other polygynous species in which nesting periods of females overlap within territories, males preferentially care for the young of their primary females (e.g., Willson 1966, Martin 1971, Alatalo et al. 1982, Johnson et al. 1993). Such a bias increases the cost of polygyny to females choosing the male after the primary females, and this increased cost can in turn affect the degree of polygyny found in the population. Such biases should apply only to nonshareable forms of parental care, as there should be no necessity to apportion shareable forms. Thus, the question of what types of parental care are shareable versus nonshareable is closely tied into the problem of apportionment. In this section we consider in turn how male red-winged blackbirds apportion their two main contributions to parental care, guarding and provisioning.

GUARDING

Guarding eggs and young is usually considered to be shareable parental care, i.e., it is assumed that a male can guard a particular brood just as well when it is one of several that he is guarding as when it is his only brood. If

38 • CHAPTER 2

so, then a male need not apportion guarding effort among his broods. As far as vigilance is concerned, Yasukawa (unpublished) found no significant difference in the mean amounts of time that male red-winged blackbirds spent attending primary (28.0%) and secondary nests (24.3%) and no significant difference in the median distances at which males guard primary (26.0 meters) and secondary nests (28.1 meters). It appears, therefore, that vigilance is shareable in red-winged blackbirds, or at least is apportioned equally. We would expect that alerting would also be shareable, because all the females on a territory should be able to hear any alarm given by the male.

In contrast, Knight and Temple (1988) presented evidence that male red-winged blackbirds apportion active defense unequally. These authors presented a crow model at the nest of the primary female while simultaneously presenting another crow at the nest of the secondary female on the same territory. The male territory owners dove at or struck the crow at the primary nest a mean of 37 times in three minutes, compared to only 11 times for the crow at the secondary nest. This difference in attack number was highly significant. This result suggests that nest guarding is apportioned unevenly and also suggests that active defense should not be considered shareable care.

Weatherhead (1990a), however, argued that the situation modeled by Knight and Temple's (1988) experiment (i.e., two predators approaching two different nests on the same territory simultaneously) is very rare. Weatherhead (1990a) assumed that the much more common circumstance would be for a single predator to menace different nests sequentially. He modeled this situation by having a human observer approach individual nests. Reaction of the male to the human "predator" was rated on a 0-to-7 scale, in which 0 meant the male departed his territory and 7 that the male struck the observer. Figure 2.4 shows the mean male defense scores graphed against day in the nesting cycle for primary and later nests. There is no bias toward defending primary nests preferentially.

Yasukawa (unpublished) found results similar to Weatherhead's using a crow model instead of humans as the predator. The model crow was presented during days 4–6 of the incubation period at one primary or one secondary nest on each experimental territory. Males responding to crows at primary nests were not more likely to attack (50%) than those defending secondary nests (58%), nor did they strike the crow more often on average at primary (22 times in five minutes) than secondary nests (19 times in five minutes). Again, there is no bias toward defending primary nests preferentially.

The discrepancy between the conclusions of these studies on the allocation of male guarding effort must in part be due to the different methods employed, specifically the use of a forced-choice paradigm by Knight and Temple (1988) versus single predator trials by Weatherhead (1990a) and

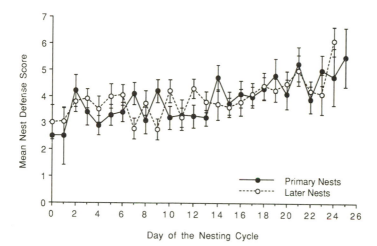

Figure 2.4 Mean nest defense scores for 52 male red-winged blackbirds at 132 nests of primary females (closed circles) and later females (open circles), graphed against day in the nesting cycle, from Weatherhead (1990a). Nest defense scores rate reaction to human observers; a score of 0 means the male left the territory, whereas 7 (the maximum) means the male struck the observer. Day 0 is the day the first egg was laid. There is no trend for defense to be more vigorous at nests of primary females than at those of later females.

Yasukawa (unpublished). Also contributing to the discrepancy may be that Knight and Temple (1988) compared defense of broods of differing ages, whereas Weatherhead (1990a) and Yasukawa (unpublished) controlled for the stage of the nesting cycle in their analyses. In Knight and Temple's (1988) experiments, predators were presented at the same time at the primary and secondary nests, which means that in most cases the primary nest would have been more advanced at the time of the trial. Several studies have found that the intensity of male defense increases as the nesting cycle advances (Searcy 1979a, Weatherhead 1990a); this trend is clearly evident in Figure 2.4. Increasing defense with increasing age of the eggs and young is compatible with "prospective" parental investment theory, which predicts that investment in the young ought to increase as the young become more costly to replace (e.g., Dawkins and Carlisle 1976, Maynard Smith 1977).

Knight and Temple (1986) have claimed that the evidence for increasing defense is a methodological artifact, caused by the defenders learning that the mock predators "respond" to their defensive efforts by "leaving" (i.e., they are removed by the experimenter) without harming the young or the defenders; thus, defense increases as a function of the number of trials measuring defense rather than as a function of stage in the nesting cycle. As evidence for their view, Knight and Temple (1986) showed that when model predators were presented only once on each territory, the intensity of nest defense did not increase through the nesting cycle. In Weatherhead's (1990a) data, however, the number of previous tests at the territory or at the nest do not explain any of the variation in defensive intensity, yet defensive intensity nevertheless increases as the nesting cycle advances. In Yasu-kawa's (unpublished) study, each male was tested only once, so repeated presentations did not confound the results.

We conclude that there is a real trend for the intensity of nest defense to increase with increasing age of the eggs and young and that this trend explains to some extent the more intense defense of primary nests when predators are presented simultaneously at primary and secondary nests. Furthermore, we agree with Weatherhead (1990a) that visits by a single predator at a time must be the rule, and simultaneous visits by different predators the rare exception. When predators do appear singly, defense by the male appears to be shareable care, there is no reason to apportion guarding unequally among broods, and males do not do so.

There is evidence, however, that males may apportion guarding according to paternity. In an unpublished study, Patrick J. Weatherhead and colleagues (pers. comm.) assessed the responses of male redwings toward "predators" (human observers) threatening broods whose paternity was determined using DNA fingerprinting. Territory owners were more aggressive when defending broods in which all young were their genetic offspring than they were in defending broods containing some extrapair young.

PROVISIONING

In contrast to guarding, provisioning is clearly not shareable: a given food item can only be provided to a single offspring, and once delivered that item is lost to individuals in any other broods the male may have. A male must therefore allocate provisioning effort among his broods in some fashion, whether he does this purely randomly, uniformly (feeding all equally), or using a more complex strategy. A number of complex strategies can be argued to be advantageous on intuitive grounds; some of these strategies have been predicted by formal models of optimal parental effort as well.

Factors to which males might respond in allocating parental care include: the size of the brood, the ages of the nestlings, the parental effort of the mate, the sex ratio of the brood, the amount of paternity, and the status of the female.

BROOD SIZE

Intuitively, one might expect favoring large broods to be advantageous, as these must be the most difficult for the female to bring off unaided and will provide the biggest payoff to the male if raised successfully. Optimization models agree that parental effort in general should increase with brood size (Winkler 1987).

NESTLING AGE

Here we might expect males to provision preferentially the most advanced broods, because these are the most likely to survive the risk of nest predation, having already passed more of the period of risk, and because early fledging young may have higher expected fitness. Greater parental effort for older young is also predicted by optimization models (Winkler 1987).

MATE'S PARENTAL EFFORT

Here one could argue either that a male's effort should decrease with an increase in the female's effort (because she is taking up the slack) or that a male should increase his effort when the female's effort increases (because the brood is becoming more valuable). Both predictions are made by optimization models, depending on whether offspring production is an accelerating function of effort (predicting male effort should increase as the female's effort increases) or a decelerating function (predicting male effort should decrease with increasing effort of the female) (Winkler 1987).

BROOD SEX RATIO

Preference might be given to broods that are predominantly male, because high-quality sons provide a higher fitness payoff than do high-quality daughters due to the greater variance in male reproductive success (Trivers and Willard 1973).

PATERNITY

Paternity refers to the proportion of young in a brood that are a male's genetic offspring. Models predict that optimal levels of parental effort should change in response to paternity given that paternity varies among matings and that males can assess paternity (Whittingham et al. 1992b, Westneat and Sherman 1993). Some models predict a continuous increase in effort with increasing paternity, whereas others predict a threshold response (Whittingham et al. 1992b).

FEMALE STATUS

It might be advantageous to prefer helping primary females, because these are older females that have demonstrated their genetic fitness and whose offspring are therefore more valuable.

Some of these hypothetical patterns were investigated by Yasukawa et al. (1990). Over a period of four years, these authors observed provisioning of 127 redwing broods on a Wisconsin upland site and subjected their results on rates of male feeding visits to stepwise multiple regression. As judged by

Table 2.2
Results of a Stepwise Multiple Regression Analysis of Male Provisioning
Rates (Visits/Hour) in a Wisconsin Upland Population.

Variable	β	SE	P
Nestling Age	0.349	0.017	<0.001
Brood Size	0.171	0.057	<0.001
Nest Status	−0.122	0.094	<0.01
Brood Sex Ratio	0.099	0.002	<0.01
Male Breeding Experience	0.095	0.042	<0.01
Constant		0.292	<0.001

SOURCE: Yasukawa et al. (1990).
NOTE: Variables are listed in order of decreasing importance in explaining variation in provisioning rates. Number of concurrently active nests and hatch date did not enter the equation at a significant level. β is the standardized regression coefficient.

the standardized regression coefficients, the most important factor influencing rates of feeding visits was the age of the nestlings in the brood (Table 2.2); males fed broods of older nestlings preferentially (Figure 2.5a). The second most important factor was brood size, with larger broods receiving more male visits (Figure 2.5b). Third in importance was nest status, defined by the order in which the first egg hatched in different broods. Rates of male feeding decreased for later nests (i.e., nests of lower status; Figure 2.5c). Of lesser importance were brood sex ratio (broods biased toward males were fed preferentially) and male breeding experience (older males performed more feeding). Once these variables had entered the regression equation, number of concurrent nests, hatch date, female breeding experience, and female feeding rate did not account for significant amounts of the remaining variation in male feeding rate. Interestingly, rates of female feeding visits showed nearly the same relationships: rates increased with increasing nestling age and brood size and decreased with declining nest status (Figures 2.5a–c). One can see that in both sexes a number of our hypothetical allocation strategies seem to be operating, with preference for older broods the most important, followed by preference for the largest brood.

In a second analysis of the same data set, Yasukawa et al. (1990) used discriminant function analysis to ask what factors determine whether or not males will provision a particular brood. This analysis looks at male feeding as a two-state variable (yes or no) rather than as a continuous one (feeding rate). Only broods that reached the eighth day after hatching were included, so nestling age was not entered as a discriminating variable. The most important discriminating variable was nest status, followed by brood size, number of concurrent nests, female breeding experience, brood sex ratio, male breeding experience, and hatch date. We mention these results because the analysis is similar to that in a second study of male provisioning, by Whittingham (1989).

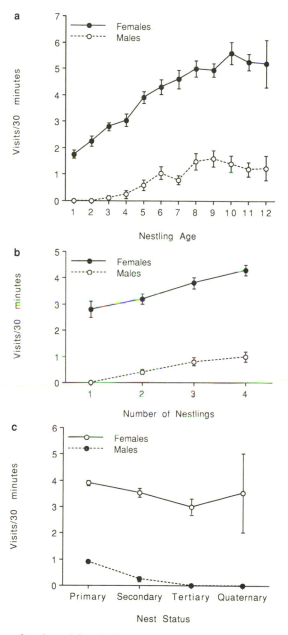

Figure 2.5 Rate of male and female feeding visits versus (a) age of nestlings in the brood, (b) size of brood, and (c) nest status, from a study of 127 broods in a Wisconsin upland, by Yasukawa et al. (1990). Nest status is defined by the order of hatching of the first eggs. Feeding rates increase with nestling age for both sexes, increase with size of the brood, and decrease with declining status. Under all circumstances, the average contribution of males to provisioning is much smaller than that of females.

In each of two years, Whittingham (1989) observed provisioning at 28 nests in a Michigan marsh. In both years, brood size was the most important factor determining whether a given nest would be aided by the male. Also important were nestling age and the date on which the eggs hatched, with males more likely to feed older nestlings and ones that hatched later in the season. In one of the two years, male provisioning was related to female feeding rate, with males more likely to provision nests at which females fed little; with this last factor, it is difficult to say which is cause and which effect. In contrast to Yasukawa et al.'s (1990) results, Whittingham found that nest status did not affect the probability of male feeding.

Whittingham (1989) confirmed the importance of brood size to male provisioning by experimentally manipulating brood size. At 12 nests where males were initially not feeding, she increased brood size from three young to five; at seven of the 12 nests the male then began feeding the young. At five nests where the male was feeding initially, brood size was reduced by removing two young (from broods of three or four); at all five, the male then ceased feeding. On two additional territories, males were initially feeding four nestlings in the primary nest and not feeding secondary broods of three nestlings. When Whittingham removed two nestlings from the primary nest, and added two nestlings to the secondary, both males switched to feeding the enlarged, secondary brood.

In an unpublished study, David F. Westneat (pers. comm.) investigated the correlation between male provisioning and paternity. Paternity was measured using DNA methods for 114 broods over four years from a mixed marsh and upland site in New York. Male provisioning rates were corrected for the usual tendency to increase with nestling age. Corrected provisioning rates showed virtually no correlation with paternity (Kendall's tau = −0.02, P > 0.50). Males fed no more at broods in which all the young were their offspring than at those broods (on their territories) in which none of the young were theirs, which belies any threshold effect of paternity on provisioning. In bivariate analyses, male provisioning was positively correlated with brood size and negatively correlated with female age, harem size, and nest order. All these correlations were weak. In a multivariate analysis, female age was the only variable significantly associated with provisioning; males provisioned more at nests of younger females.

The various studies of male provisioning do not agree in all respects on what factors are important in determining male allocation of feeding effort. There is, however, a simple underlying mechanism that explains many of the observed patterns: male response to nestling hunger. Those factors that affect male provisioning are, for the most part, ones that are obviously related to the food requirements of nestlings and thus to their hunger. Food requirements increase with nestling age, and males prefer to feed older nestlings. Food requirements increase with brood size, and males prefer to provision larger broods. The less the female provisions, the hungrier the nestlings are, so male provisioning increases when female provisioning decreases. Young females are probably less adept at finding food to take to young, so their broods are hungrier and males provision them more. Conversely, factors unrelated to food requirements or hunger of nestlings, such as paternity, do not affect male provisioning.

The importance of nestling hunger to male provisioning has been demonstrated experimentally by Whittingham and Robertson (1993). These authors removed three nestlings from broods of four on territories of males that had not been feeding young, and deprived the removed nestlings of food for two hours. When the nestlings were placed back in their nests, they begged constantly. In response, nine of 11 males began provisioning, compared to none of 11 males that provisioned at matching control nests. Conversely, when the experimenters fed nestlings to satiation with mealworms, the nestlings ceased begging and the males temporarily ceased provisioning. Additional evidence is provided by experiments of Julie Burford and Teresa J. Friedrich (unpublished), performed on Yasukawa's Wisconsin population. Burford and Friedrich placed loudspeakers near redwing nests and broadcast five minutes of either nestling begging calls or white noise. Six of 28 males began feeding nestlings immediately following playback of the begging calls, whereas none of 13 males provisioned following playback of noise.

Major studies of male provisioning do not agree on the importance of nest order to male provisioning: Whittingham (1989) found no relationship between nest order and male provisioning, David F. Westneat (pers. comm.) found a negative relationship that disappeared when other factors were controlled statistically, and Yasukawa et al. (1990) found a negative relationship that held good despite statistical control of other factors. The differences in these results may be due in part to differences in the overlap of the nesting attempts of females sharing a male's territory. Females within harems overlapped substantially in timing of nesting in Yasukawa et al.'s study, but not in Whittingham's. Males in Yasukawa et al.'s study were thus faced with choosing among simultaneously active nesting attempts, a choice not faced by males in Whittingham's study. In fact, when Yasukawa et al. (1990) examined cases in which one of their males fed nestlings of more than one brood, they discovered that the broods were usually separated in time. In such cases, earlier nests were not more likely than later ones to be provisioned by males, and the results were very similar to those of Whittingham (1989).

In any case, there are probably conflicting forces acting on the likelihood of male's feeding nests of different status. On the one hand, males may be selected to favor nests of high status (i.e., that hatch relatively early) because these contain the oldest and largest broods. On the other hand, males may also be selected to delay feeding until resident females have completed their clutches and are no longer fertile (Whittingham 1994), which will tend to reduce the probability that nests of high status will be fed. These opposing forces may balance differently in different areas, producing different patterns of male provisioning versus nest status.

It is clear that the likelihood of male provisioning is influenced both by factors that have little relationship to female status, such as brood size and sex ratio, and by at least one factor that is closely tied to female status, namely nest status. All these factors work together to produce the pattern by which males allocate their provisioning among females of different status, and it is this pattern which is important to the evolution of polygyny. Data on the observed allocation pattern, garnered from various studies, are presented in Table 2.3. It should be noted that most of the studies cited in this table used nest status to assess female status, under the assumption that females begin nesting in the same order that they settle, whereas in fact the correlation between nest status and female status is known to be positive but imperfect (Lenington 1980, Teather et al. 1988). The exception is the study by Searcy (unpublished), which directly assessed female settlement. It is clear from Table 2.3 that in some studies males favor their primary broods

Table 2.3
Percentage of Females of Differing Status Whose Broods are Fed by Males.

Location	Female Status	N	% Fed by Males	Reference
Indiana	primary & monogamous	46	50	Patterson 1979
	secondary	37	8	
	tertiary +	36	3	
Ontario	primary & monogamous	29	62	Muldal et al. 1986
	secondary	23	74	
	tertiary	11	36	
	quaternary +	8	0	
Michigan	primary	10	60	Whittingham 1989
	secondary	4	75	
	teritary +	3	67	
Wisconsin	primary & monogamous	233	63	Yasukawa unpublished
	secondary	113	34	
	tertiary	32	25	
	quaternary	10	0	
Pennsylvania	primary	15	47	Searcy unpublished
	secondary	9	44	
	tertiary +	7	43	

over later ones (Patterson 1991, Yasukawa unpublished). In other studies, the allocation of male help among females of differing status is quite uniform (e.g., Whittingham 1989, Searcy unpublished).

PROVISIONING VERSUS GUARDING

Finally, it is worth considering whether the two forms of male parental care compete with each other, and the extent to which trade-offs between them affect the cost to females of sharing a mate. Guarding requires male attendance on the territory, whereas provisioning males often must forage away from the territory. What happens to a male's ability to guard when he is provisioning? Is guarding at nests of secondary females disproportionately affected by the male's decision to feed primary nestlings?

One way to examine the possible effects of trading off between guarding and provisioning is to compare male attendance at nests whose broods the males provision with attendance at nests of unprovisioned broods. We make this comparison with data from Newark Road Prairie, restricting analysis to primary broods, which are the ones most frequently provisioned at that locality (Table 2.3). As shown in Figure 2.6, the decision to feed nestlings means that males spent considerably less time in attendance (i.e., perched within 10 meters of the nest). Interestingly, the difference occurred long

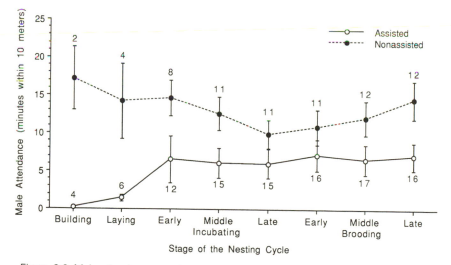

Figure 2.6 Male attendance at primary nests for males that did (N = 15) or did not (N = 34) assist females in provisioning nestlings, from a study of an upland Wisconsin site by Yasukawa (unpublished). Assisting males attended nests less than those not assisting, starting long before any provisioning occurred.

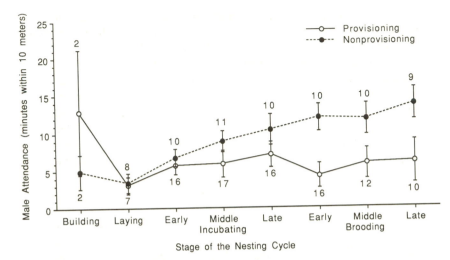

Figure 2.7 Male attendance at secondary nests for males that did (N = 11) or did not (N = 22) provision nestlings of primary nests, from a study of an upland Wisconsin site by Yasukawa (unpublished). Males that were provisioning their primary broods guarded their secondary nests less than males that were not provisioning their primary broods.

before provisioning actually began in the midbrooding period. We might speculate that males trade guarding against provisioning and make their decisions very early in the nesting cycle, but much more information is clearly needed.

A second way to examine trading off is to limit our analysis to secondary nests, which are less frequently provisioned at Newark Road Prairie, and compare attendance by males who provision their primary nests with that of nonprovisioning males. Figure 2.7 shows that secondary nests were attended far less when males were busy provisioning primary ones than when they were not so engaged. This difference was most pronounced late in the nesting cycle.

These two analyses show that provisioning and guarding are incompatible activities, at least to some extent. Males must therefore partition their parental care between these two components, and the decision rules may differ depending on the female's status, at least in some populations. It will be interesting to see whether the Newark Road Prairie pattern of partitioning occurs in other habitats and geographic locations.

Conclusions

Female red-winged blackbirds probably provide more parental care per nest than do males, but males nevertheless provide considerable assistance in the form of nest guarding and provisioning. Male nest guarding seems to be, for the most part, shareable care, whereas provisioning is definitely nonshareable. Therefore, females choosing an already-mated male will pay a cost in having to share his provisioning effort, but not through sharing his nest guarding. The cost of sharing male provisioning must, on average, be greater in eastern populations, where male provisioning is common, than in western populations, where it is uncommon. The cost of sharing male provisioning also depends on how males apportion feeding among females of differing status. The pattern of apportionment appears to be variable, with primary females favored at times, while at others male provisioning is apportioned evenly.

Males respond to many factors, such as nestling age, brood size, and nest order, in allocating their provisioning effort among broods, but they do not respond to paternity. Males do seem, however, to allocate nest guarding according to paternity, guarding broods consisting of their own genetic offspring most intensely. That males should respond to paternity in allocating one form of parental care and not the other seems paradoxical; we would expect both forms to be allocated if males could assess paternity, and neither to be allocated if they could not. This situation certainly merits further study. At any rate, if males do allocate either form of parental care based on paternity, this should select against the performance of EPCs by females.

Several other questions about parental care remain unanswered, despite considerable interest in this topic. We have focused on allocation decisions of males among broods, but because a brood typically contains more than one nestling, additional decisions must be made within broods. Do males allocate nonshareable investment randomly among nestlings within a brood, and if not, what rules do they follow? Perhaps the most intriguing problems, however, concern the general geographic pattern of male feeding. Is the east-west difference in the frequency of male feeding real? If so, does it have a genetic basis? Why do males of some populations feed frequently and those of others feed infrequently? What are the fitness costs and benefits that determine whether male feeding is advantageous? The variation in male feeding that has been observed over shorter distances and over time is also intriguing. What is the basis of such variance and is it adaptive? Finally, there is a clear need to extend studies of parental care to the end of the period of dependence. What are the patterns of paternal and maternal care provided to fledglings? What factors affect the male's allocation of care to fledglings? These and other questions are likely to motivate continued research on parental care in red-winged blackbirds.

3   Territoriality

The mating system of red-winged blackbirds is territorial polygyny, a type of polygyny in which the prolonged association between one male and several females occurs on the male's territory. As is typical in territorial polygyny, male red-winged blackbirds establish their territories first, and females settle on territories later (for an interesting exception, see Davies 1992). Female red-winged blackbirds make extensive use of the resources provided by the territory, and there is good evidence that their choice of mate is in large part determined by the quality of those resources (see chapter 5). By affecting female choice, variation in territory quality may have an important effect on the degree of polygyny shown by a population. Territoriality, then, is a basic component of social polygyny in red-winged blackbirds and must be understood if we are to understand the species' mating system. As we shall see later (chapter 7), acquisition and continued ownership of a territory are crucial to a male's mating and reproductive success, so understanding territorialty is also crucial to understanding sexual selection.

Male red-winged blackbirds expend considerable amounts of time and energy in territory defense, presumably because of the importance of the territory to their reproductive success. Territory defense is compatible with some other behaviors important to males, such as guarding mates against attempts by other males to copulate with them; in fact over parts of the breeding season, territory defense and mate guarding may be considered identical behaviors. Territory defense is less compatible, however, with other important behaviors, such as parental care and the pursuit of extrapair copulations. Understanding territoriality is therefore also important to understanding the balance of costs and benefits of these other behaviors.

In this chapter we explore some basic aspects of the territorial system of male red-winged blackbirds. We first describe the types of habitats defended and give some information on areas of territories and the duration of defense. Next we describe the displays and other behavior used by male redwings in territory defense and discuss how well their spacing behavior satisfies various definitions of territoriality. We then discuss several measures of the intensity of territoriality, such as time budgets, pressure from males without territories, and amount and ferocity of fighting. Finally we discuss site fidelity of returning territory owners and acquisition of territory by new owners. We postpone consideration of some other aspects of territoriality, in particular of the factors leading to success in competition for territory, until later (see chapters 7 and 8).

Where and When Males Are Territorial

Throughout their range, male red-winged blackbirds seem to prefer holding territories in marshes, particularly marshes dominated by cattails. In many regions, any patch of cattails comprising more than a few tens of square meters is defended. For example, a male in Indiana, who had the misfortune to live next to Yasukawa's house, defended a territory with a patch of cattails comprising a mere two square meters and managed to acquire two females who nested in this tiny patch. Other types of marsh vegetation are also defended, including bulrushes, bur reed, buttonbush, willows, various cord grass species, and purple loosestrife. In Costa Rica, Orians (1973) found red-winged blackbirds defending territories in marshes dominated by bunch grasses and sedges as well as cattails. Territories on the margins of marshes often include dry areas, containing grassy areas used for feeding or trees used as display perches.

Male red-winged blackbirds may also defend territories in upland areas away from marshes. This habit is more prevalent in eastern and midwestern North America, where males quite commonly hold territories in upland hayfields and grassland, in old fields in the early stages of succession, and even in open patches in woodland (Bent 1958, Robertson 1972, Case and Hewitt 1963). Defense of upland territories is more rare in western North America, but it does occur. For example, male redwings in California were observed holding territories on an upland site dominated by wild radish and mustard (Ritschel 1985), and Bent (1958) gives a record of male redwings defending territories in willows on the banks of a canal, again in California.

Table 3.1 shows areas of territories from a number of locales. There is obviously a great deal of variation in territory sizes, both within and between sites. Some of the between-site variation is explained by habitat, with territories in general being smaller in marshes than uplands. The overall average is approximately 2000 square meters, which is considerably smaller than for most birds of comparable body size. Schoener (1968) calculated a regression of log territory size on log body weight for 61 species of birds in which both males and females feed principally on the territory. Assuming a mean size of 70 grams for male redwings, Schoener's curve predicts a territory size of about 20 acres, or 81,000 square meters. It could be argued that Schoener's curve should not be used for red-winged blackbirds, because both sexes in this species often feed much of the time off the territory; however, we believe that the habit of feeding off the territory is more a consequence than a cause of the small territory sizes in redwings.

Table 3.1
Territory Sizes (in Square Meters) of Male Red-winged Blackbirds.

Mean Area	Range	N	Locale	Habitat	Reference
330	124–583	17	Wisconsin	marsh	Nero 1956b
762	—	86	California	marsh	Orians 1961
688	242–4532	51	New York	marsh	Case & Hewitt 1963
2185	121–4006	49	New York	upland	Case & Hewitt 1963
476	71–1607	89	Washington	marsh	Holm 1973
1810	487–4343	16	Costa Rica	marsh	Orians 1973
3210	700–9600	10	Iowa	upland	Blakley 1976
1045	153–2890	97	Ontario	marsh	Weatherhead & Robertson 1977a
810	—	68	California	marsh	Ritschel 1985
370	—	13	California	upland	Ritschel 1985
5816	2214–29,235	—	Ontario	upland	Eckert & Weatherhead 1987c
8319	1188–18,456	95	British Columbia	marsh	Picman 1987
384	266–556	6	Alberta	marsh	Dickinson & Lein 1987

The seasonal duration of territory defense varies with latitude, being shorter in the North, as one would expect. Near the northern limit of the breeding range, in Saskatoon, Saskatchewan (latitude 52°N), the first territorial males arrived on average on April 16 (Miller 1968). The last young fledged early in July, and presumably the last males abandoned their territories at this time also. In central California (latitude 43°N), Orians (1961) found that the first males defended territories in early January, though up until March they remained on territory only for brief periods in the early morning and late afternoon. These males gave up their territories about the time that females finished nesting, in late June. At the southern limit of the species range, in Costa Rica (latitude 10°N), Orians (1973) found males defending territory throughout the year except at the height of molting during September and October. Thus the general trend is for the duration of defense to increase from a low of less than three months at the northern end of the species' range to a high of more than 10 months at the southern end. Exceptions to this trend can be found, however; for example, males in Washington State (latitude 46°N) have recently begun to defend territories nearly year-round (Beletsky and Orians 1987a), apparently because human agricultural practices make it possible for males to obtain food near their territories during the winter.

Visual and Vocal Displays

In maintaining a territory, a male red-winged blackbird's first line of defense is a variety of visual and vocal displays, which apparently serve to announce the displayer's ownership of the territory and to discourage other males from entering it. One or two of these displays function only in defense, but most function in both defense and courtship. There are also a few displays that function only in courtship. Below we describe all these displays together. We also describe what may happen when displays are not sufficient to repel an intruder, and a territory owner has to resort to escalated defensive behaviors.

Song

The display with the longest effective range is the song. To unbiased observers (i.e., not us), redwing song may seem an unmelodious vocalization; it is often described as "conc-a-reeee." A spectrogram of a song is shown in Figure 3.1, with the constituent parts labeled. The first third of the song, the "conc-a" part, consists of a jumble of unrepeated introductory notes. The remaining two-thirds, the "reeee," usually consists of a single syllable repeated in a steady-rate trill. Sometimes one or more unrepeated terminal notes follow the trill. Very occasionally a song will be "double," with the

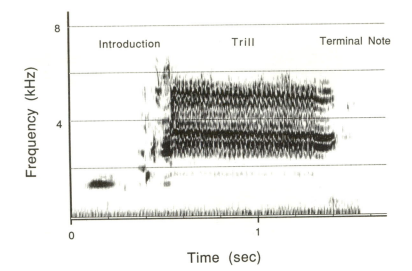

Figure 3.1 Spectrogram of the song of a male red-winged blackbird recorded in Pennsylvania. The song consists of two major parts, a jumble of unrepeated introductory
notes ("conc-a") followed by a trill consisting of a single, rapidly repeated syllable
("reeee"). Some songs, such as this one, also have one or more short, post-trill notes.

trill part given twice. Songs vary in length from approximately 0.7 to 1.75
seconds, with the average about 1.2 seconds (Orians and Christman 1968,
Yasukawa 1981b). Energy is concentrated between 900 and 6000 hertz.
There is a great deal of variability in all parts of the song, both within
individuals, among individuals within populations, and among populations.

Within individuals, songs can be classified into a discrete number of
"song types," each of which is a markedly different version of the species
song. A classification of songs into song types accounts for much of the
within-individual variability, but not all; there is some remaining within-
song type variability, i.e., different renditions of the same song type can
differ from one another in small details. The collection of song types sung
by one male is termed his "song repertoire"; one such repertoire is illustrated
in Figure 3.2. Males in Indiana and New York have between two and eight
song types in their repertoires, with a mean repertoire size of 4.3 (Yasukawa
1981b). Males in Florida and California also have repertoires of about this
size (Kroodsma and James 1994).

Male redwings sing in a pattern termed "eventual variety," repeating a
single song type many times before switching to another. Yasukawa (1981b)
found a mean of 19.8 repetitions between switches, though occasionally a
male would sing a bout of more than 100 renditions of a single song type.
The song types within a single male's repertoire appear to be as different
from one another as song types taken from different repertoires, both in

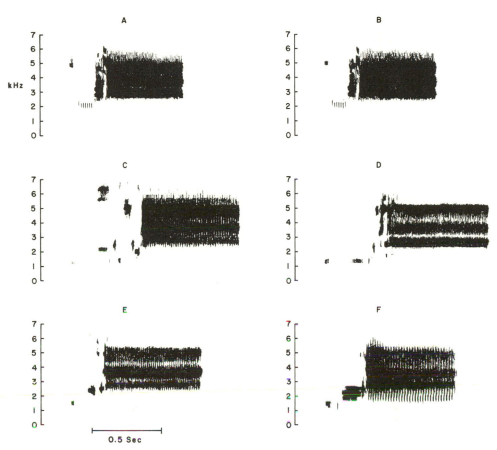

Figure 3.2 Song repertoire of a male red-winged blackbird, recorded in New York, from Searcy and Yasukawa (1983). The top two sonograms (A and B) show two different renditions of the same song type. The other four sonograms (C–F) show the other four song types in this male's repertoire.

terms of acoustic measurements (Yasukawa 1981b) and as judged by redwings themselves (Searcy et al. in press). Occasionally, song types sung by different males in the same population will sound similar, but we know of no examples where two such song types have actually turned out to be identical when analyzed sprectrographically; therefore, we regard each song type of each male as unique. Songs function both in territory defense and courtship (see chapter 8).

CALLS

Male red-winged blackbirds give a number of vocalizations classified as calls. Included are a variety of alarm calls, which we have already described (chapter 2). Two other calls are the growl and precopulatory note. Growls are short, harsh vocalizations given in both aggressive and courtship contexts. Precopulatory notes are single, short notes, typically produced in series and only during courtship. Their form differs between males, even within populations.

Song Spread

Closely associated with song is a postural display called the "song spread." Song spread is used both in territory defense and courtship and is the most frequent display posture given by territorial males. Typically, a male will give song spreads at rates of 3 to 5 per minute during the early part of the breeding season. The song spread exposes and exaggerates the epaulets, for which the red-winged blackbird is named, as the male delivers a song. When a male is in a relaxed posture, the red part of the epaulet is completely covered by the black over-wing coverts, so that only the narrow, yellow margin of the epaulet is visible. (In the Central Valley of California, where the redwing is sympatric with the tricolored blackbird, the yellow margin is missing and a relaxed male appears entirely black.) During song spread, the coverts are retracted to expose the red feathers, and the epaulet feathers may be erected and vibrated to increase their conspicuousness.

Song-spread display shows extreme variation in intensity, and the more intense the song spread, the more conspicuously are the epaulets displayed. Peek (1972) arbitrarily divided what is really a smooth continuum of song-spread intensities into four categories: (1) incipient, in which the red portion of the epaulet is exposed, the head is thrust forward slightly, and the tail is spread, while the wings remain folded against the body; (2) low intensity, in which the epaulets are exposed, the tail is lowered and fanned more fully, and the wings are moved away from the body slightly; (3) moderate intensity, in which the epaulets are exposed and raised, the tail is lowered and fanned, and the wings are moved well away from the body in a flat plane; and (4) high intensity, in which the epaulets are raised and vibrated, the tail is fully fanned, and the wings are extended fully and bowed downward, so that the bird assumes a disc-like shape (Figure 3.3).

Flight Displays

Male red-winged blackbirds have a number of displays that are given in flight, and in all of them the red-and-yellow epaulets are conspicuous. In normal flight, the epaulets are exposed by the movements of the wings, so even though such flight is not a specialized display, it still serves to identify the male and advertise his location. In "flight song" (Figure 3.4), most contour feathers and the epaulets are erected, and the tail is lowered and spread as the male flies slowly or glides over his territory (Orians and Christman

Figure 3.3 Male red-winged blackbird in a high-intensity song-spread display, from Orians and Christman (1968). The wings are extended and bowed downward, the epaulets are flared, and the tail is fanned while the male sings. This display is used both in defense of territory against other males and in attraction and courtship of females. Reprinted with permission from G. H. Orians and G. M. Christman. *A Comparative Study of Red-winged, Tricolored, and Yellow-headed Blackbirds.* Copyright 1968 by the Regents of the University of California.

1968). The male usually sings while gliding. Erection of the epaulets increases their visibility, whereas the slow movement of the male increases the time he is above the vegetation and thus conspicuous. Typically, flight song is elicited by the approach of another male or of a female. A second flight display is "fluttering flight," in which the epaulets are flared while the male

Figure 3.4 Male red-winged blackbird giving a flight-song display, from Orians and Christman (1968). As its name implies, this display is given in flight while the male sings. Note that the tail is lowered and spread while the epaulets are flared. Flight-song is used in both defense of territory and attraction of females. Reprinted with permission from G. H. Orians and G. M. Christman. *A Comparative Study of Red-winged, Tricolored, and Yellow-headed Blackbirds.* Copyright 1968 by the Regents of the University of California.

Figure 3.5 Male red-winged blackbird giving a bill-up display, from Orians and Christman (1968). The bill is pointed up, the epaulets are exposed, and the contour feathers are sleeked. This is an aggressive display used in male-male interactions, typically when the interactants are quite close to each other. Reprinted with permission from G. H. Orians and G. M. Christman. *A Comparative Study of Red-winged, Tricolored, and Yellow-headed Blackbirds.* Copyright 1968 by the Regents of the University of California.

flies with rapid but shallow wing strokes (Orians and Christman 1968). Fluttering flight is most common early in the breeding season, when territories are being established and females are settling.

Although flight displays are often visually spectacular, they are relatively rare in comparison with perched displays, perhaps because they are more expensive energetically. Flight displays probably function both in territory defense and attraction of females.

BILL-UP

The bill-up is a display given by perched males and seems to be associated with the establishment and defense of territorial boundaries. In a bill-up, the male sleeks his contour feathers, exposes his epaulets, stretches his neck, and points his bill skywards at a 45 to 90° angle above the horizontal (Orians and Christman 1968; see Figure 3.5). This display is frequently given by males near the boundaries of their territories, and often two territorial neighbors will give bill-ups to each other across their mutual boundary. Bill-up displays are also often given by territorial males in response to trespassing males, and usually such display is sufficient to cause a trespasser to leave the territory without further interaction. Females also give bill-up displays in response to other females, especially during the early portion of the breeding season when prospecting females attempt to settle on territories that already have resident females (Nero 1956a, Orians and Christman 1968). The bill-up thus seems to function in intrasexual contexts as an aggressive display.

CROUCH AND PRECOPULATORY DISPLAY

In the crouch, a perched or standing male flexes his legs, thus lowering his body, while bowing his head even lower. The tail is spread and lowered, the

Figure 3.6 Male red-winged blackbird in precopulatory display, from Orians and Christman (1968). The male crouches, with his head bowed, his legs flexed, and his wings alternately drooped and fluttered rapidly. This display is given in courtship. Reprinted with permission from G. H. Orians and G. M. Christman. *A Comparative Study of Red-winged, Tricolored, and Yellow-headed Blackbirds*. Copyright 1968 by the Regents of the University of California.

wings are extended and drooped, and the epaulets are flared (Orians and Christman 1968). An intense crouch becomes a precopulatory display if the male adds a rapid fluttering of the wings (Figure 3.6) together with pre-copulatory calls. Typically, a male in precopulatory display walks forward in a convoluted path, pausing occasionally to sing. Crouch is usually given during courtship of a receptive female but is occasionally given in aggressive situations to other males (Orians and Christman 1968). Precopulatory display is given exclusively in courtship. Again, both these postures display and exaggerate the epaulets.

Escalated Defense

Display alone is often sufficient to cause an intruder to leave, but if not, the owner can escalate by flying at the intruder and attempting to supplant him. Sometimes an owner will go directly to this tactic, without using elevated display. An intruder usually gives way before an owner, moving enough to avoid him and often leaving the territory altogether. Much more rarely, the intruder refuses to give way, and a fight results. Fights may start with the combatants on the ground or in the air. When fighting on the ground, two males will tumble about, grasping each other's plumage with their feet, pecking at each other with their sharp bills, and striking each other with the wrists of their wings (Orians and Christman 1968). Birds fighting in the air fly up facing each other and striking out with feet, bills, and wings (Orians and Christman 1968); often they grapple and then fall, locked together, to the ground and continue fighting there. Yasukawa observed one such encounter in which the males continued to fight even after they fell into the water of Yellowwood Lake.

Definitions of Territoriality

Thus far we have been asserting that male red-winged blackbirds are territorial without considering what this really means. "Territoriality" has been defined in a variety of ways, and the term has been applied to some rather different spacing systems. Certain definitions of territoriality employ a behavioral criterion; that is, they specify how the owner must behave in order to be considered territorial. Specifically, these definitions require that the owner show space-centered advertisement and aggression, as in Noble's (1939) definition of a territory as "any defended area." Operationally, an individual is considered to be territorial if its aggressive display and/or overt aggression are concentrated within a circumscribed area. A second class of definitions stress an ecological criterion, namely that the owner has exclusive access to an area (Pitelka 1959). To satisfy these definitions, one needs to show that there is little or no overlap in use of space by different individuals (Waser and Wiley 1979). A third type of definition, which we favor, requires that both the behavioral and ecological criteria be met. Here an example is Wilson's (1975) definition of a territory as "an area occupied more or less exclusively by an animal or group of animals by means of repulsion through overt defense or advertisement."

We can test the behavioral criterion for male red-winged blackbirds using data gathered by Searcy (1986) on a cattail marsh in western Pennsylvania. Lines of numbered poles were set out at five-meter intervals to form one X-axis and four Y-axes. Using the resulting grid, observers could specify the locations of males within five-meter squares. Three color-banded males were studied throughout the period of female settlement and nesting. Figure 3.7 shows least convex polygons drawn around the locations of songs for each male. Obviously, display is given from circumscribed areas by all three males, as the behavioral definitions require. For two of the males, a similar analysis can be performed on locations of aggressive acts, locations where the focal male attacked or directed a bill-up at another male (Figure 3.8). Aggressive acts are also clearly confined to discrete areas, as required.

Data on these same males can be used to test the ecological criterion, as to whether individuals have exclusive access to particular areas. One way to

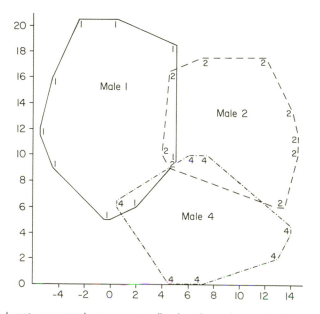

Figure 3.7 Least convex polygons surrounding locations of songs for three males on a marsh in Pennsylvania, from Searcy (1986). The three males were observed during the periods of female settlement and nesting, and their locations when singing were noted in terms of 5-×-5-meter grid squares. Sample sizes were 1035 songs for Male 1; 940 for Male 2; and 218 for Male 4. The resulting "areas defended by song" are highly exclusive.

assess exclusive access is to measure the amount of overlap between areas defended through display and aggression by adjacent males. Searcy (1986) calculated the overlap of male A on male B as the area shared by males A and B divided by the total area defended by B. Table 3.2 gives overlap values for areas defended by song and for areas defended by aggression. All values are less than 0.10. Clearly, defended areas are fairly exclusive with regard to neighboring residents, but we should also consider exclusivity with regard to nonterritorial males. Intrusions by nonterritorial males are fairly common, but they are usually brief, so that territories are free of such intruders most of the time. For the three males in the Searcy (1986) study, territories were free of nonterritorial intruders roughly 98 to 99% of the time, depending on the male (Searcy unpublished).

The territories of male red-winged blackbirds, then, meet both the behavioral and the ecological criteria, and thus they also meet definitions such as Wilson's (1975), which specify both criteria. Male redwings, then, are clearly and classically territorial.

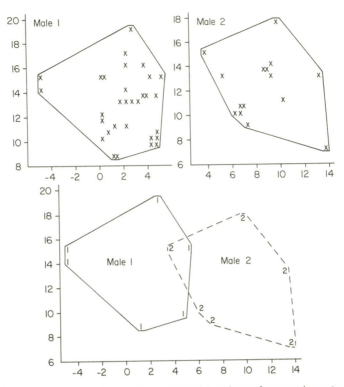

Figure 3.8 Least convex polygons drawn around locations of aggressive acts for two male red-winged blackbirds on a marsh in Pennsylvania, from Searcy (1986). Each x marks a 5-×-5-meter square in which the focal male flew at or pecked another male or performed a bill-up display. Note that the "areas defended by aggression" are largely exclusive.

Intensity of Competition for Territories

One way to gauge the intensity of competition for territory is to examine the time and energy budgets of territorial males. This has been done for red-winged blackbirds by Orians (1961), who measured time budgets for several males defending territories on a cattail marsh in California. Time spent by males on territory was very low when defense began in midwinter but then rose to about three-fourths of the daylight hours at the height of breeding. Orians estimated that about one-fourth of the time on the territory, or three-sixteenths of total daylight, was spent in territory defense, specifically in vocalizing, postural displays, chasing, and fighting. On the one hand, this fraction may overestimate the true investment in territory defense, in that some of the territorial displays function in female attraction as well as in

Table 3.2
Overlap in Areas Defended by Song or Aggression, for Males on a
Marsh in Pennsylvania.

| Overlap By | Areas Defended by Song | | | Overlap By | Areas Defended by Aggression | |
| | Overlap On | | | | Overlap On | |
	m1	m2	m4		m1	m2
m1	—	0.062	0.042	m1	—	0.066
m2	0.047	—	0.092	m2	0.065	—
m4	0.030	0.084	—			

SOURCE: Searcy (1986).

defense. On the other hand, investment in defense may be underestimated in that periods spent quietly surveying the territory from a perch were not included, though during such periods males can watch for intruders.

Yasukawa (unpublished) also constructed time budgets, by observing males during the morning hours on his Indiana marsh. Figure 3.9 shows that males were on their territories 75 to 95% of the time during mornings

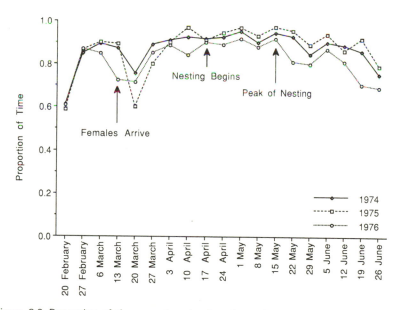

Figure 3.9 Proportion of time spent on territory by territory-owning male red-winged blackbirds at Yellowwood Lake, Indiana. Observations were made during mornings in 1974–1976. The data shown for each year are means for 22, 23, and 28 males, respectively. Males on average spent 75 to 95% of their time on territory throughout the breeding season.

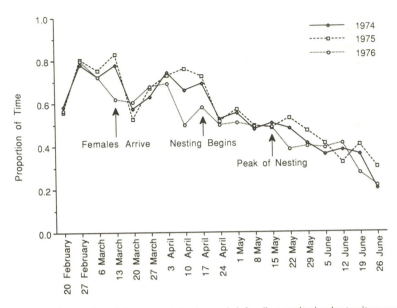

Figure 3.10 Proportion of time spent singing and defending territories by territory-owning male red-winged blackbirds at Yellowwood Lake, Indiana. Observations were made during mornings in 1974–1976. The data shown for each year are means for 22, 23, and 28 males, respectively. At peak periods, males spent on average 60 to 80% of their mornings engaged in activities related to territory defense.

throughout the breeding season. Time spent on territory was particularly high during the period when females were nesting. Figure 3.10 shows the proportion of time spent by males in activities that can be considered territory defense, such as watching over the territory from an exposed perch, singing, chasing intruders, etc. Time spent in defense peaked at 60 to 80% of the morning hours early in the season, around or just before the time females began arriving at the marsh, and then declined through the nesting season. Again, one can argue about what activities to include as territorial, but it is nevertheless clear that male red-winged blackbirds make a very substantial time investment in territory defense.

Estimates of energy budgets are even more problematic, as we do not know the energy cost per time of activities such as song spreads and bill-ups. Given the substantial time devoted by male redwings to territory defense activities, and the fact that some of these activities (such as chasing intruders) must be quite expensive in energy, it is clear that males also make a substantial energy investment in defense.

A second approach to gauging the intensity of competition for territory is to consider the amount of pressure from nonterritorial males. It has long been apparent that there are large numbers of such floating males in redwing populations and that many of these males strive to enter the ranks of territory owners. Orians (1961) attempted to assess the numbers of such floaters by shooting territory owners on small sections of two large marshes in California. Owners were removed on five days spaced through a breeding season on one marsh, and on eight days on the second marsh. New males replaced removed ones quickly enough so that the pool of owners present for removal was never exhausted. Subadult males began showing up among the replacement males as removals continued, but throughout the season the bulk of replacements were adults. Males were not banded, so Orians could not say if the replacements were neighbors, floaters, or territorial males from elsewhere.

More information on the source of replacements has been obtained in subsequent removal experiments. Beletsky and Orians (1987b) removed owners from 55 territories in Washington State. Forty-three of the territories were taken over by one or more neighbors, and the remaining 12 territories were taken by 15 new replacement males. Seven of this latter group were previously banded males of known history: three of these were males who had been removed from other territories and then released, and the other four were males never known to have held a territory previously. The remaining eight replacements were unbanded adults; as nearly all the territory owners in the population were banded, these eight were almost certainly from the floating population.

Males were removed over a two-year period from 41 territories in Ontario; all 41 territories were reoccupied within the same breeding season, by a total of 44 males (Eckert and Weatherhead 1987d). Removed males were held through the experimental period, so none of the replacements could have been males removed from other territories. All the territories in this experiment were on isolated "pocket marshes," each defended by a single male without neighbors, so replacements were not neighbors either. Given the site fidelity of territory owners (see below), it seems unlikely that replacements were territory owners shifting territories. Thus all 44 replacements in this experiment were most probably from the nonterritorial population.

Another sign of the amount of pressure on territory owners is the speed with which replacement can occur. Orians (1961) noted that some replacements occurred within an hour of removal in his experiments, and one within 15 minutes. Eckert and Weatherhead (1987d) found that 28 of 41 vacant territories were reoccupied within a day of removal. Other authors have noted that naturally-disappearing males are often replaced equally rapidly (e.g., Nero 1956b). One of us once captured an owner as a demonstration for a group of visiting graduate students; this male was banded and measured in rather more leisurely fashion than usual, and during the 10 minutes or so he was away from his territory, an intruder took his place.

The data from removal experiments indicate there are many floaters present but do not allow us to estimate the ratio of floaters to owners. Payne (1979) attempted to census floating males as well as territory owners on a wetland site in Michigan. During each of two years, territory owners and nesting females were counted, and floating males were captured, banded, and released. A mean of 15 territorial males and 34 females were present per year, compared to a mean of 16 adult floaters and 38 yearling floaters. Taken at face value, these data indicate there were approximately one adult and 2.5 subadult floaters per territory owner. The problem with these estimates is that we do not know the size of the area from which floaters were drawn for capture; if this area is larger than the area over which owners were counted, then the ratio of floaters to adults would be overestimated. Unfortunately, little is known about the behavior of floating males in red-winged blackbirds, in particular about the size of their home ranges.

Some idea of the relative numbers of floating and territorial males can be gained by comparing sex ratios among all adults to sex ratios among breeding adults. Surveys of mixed-sexed flocks made out of the breeding season typically find more males than females one year old or older (Meanley 1964, Payne 1969), but these results may be biased by the greater conspicuousness of males. It is probably safer to infer adult sex ratios from nestling sex ratios and relative survival of the sexes. The sex ratio at hatching seems to be 1:1. Weatherhead (1983) found a sex ratio of 1:0.94 (male:female) in 97 clutches that experienced no egg loss or infertility, and Weatherhead (1985) found that 46% of 554 eggs hatched into males. Weatherhead and Teather (1991) reviewed data from several studies on sex ratio among fledglings and found a mean ratio of 0.894:1 (male:female), with a range of 0.648:1 to 1.209:1. Using national banding data, Searcy and Yasukawa (1981a) obtained estimates of male and female survival from the age of one year on using two different methods; results indicated that survival of males was as high as for females. No good estimates of relative survival between fledging and the age of one exist, but if we assume that survival is roughly equal for males and females during this period, then the overall sex ratio for redwings aged one year or older ought to be close to that at fledging, i.e., approximately 0.9:1. It is thought that all female redwings aged one year or older breed, the evidence being that all females captured near the end of the breeding season show signs of a brood patch (Holcomb 1974).

If there are 0.9 times as many males as females aged one or older, and all the females breed, then it follows that the number of males aged one or older equals 0.9 times the number of breeding females. This allows us to estimate the number of nonterritorial males from mean harem sizes. For example, if the mean harem size in a population is three, then there must be 2.7 males per harem (3 × 0.9 = 2.7) and 1.7 floaters per territory owner (2.7 − 1 = 1.7). By this reasoning, the various redwing populations for which harem sizes have been estimated (Table 1.1) would have from 0.5 to

4.6 nonterritorial males per territory owner. The overall mean harem size from Table 1.1 is 3.2, which yields an overall estimate of 1.9 floaters per owner. These figures must be inexact, but they do convey a conclusion that is undoubtedly correct, namely that a very high proportion of the male population lacks territories.

Many of the nonterritorial males are, of course, one-year-old subadults. If we assume that survival is 54% per year for males (Searcy and Yasukawa 1981a) and is constant from the age of one year on, as is usually approximately true for birds of this size (Deevey 1947, Ricklefs 1973), then subadults would make up 46% of the male population one year old and older. Assuming that almost all one-year-olds are floaters rather than owners, one-year-olds would make up about 70% of the floater population. This still leaves on average approximately 0.6 adult floaters per owner. As subadult males are usually subordinate to adults (Wiley and Harnett 1976, Searcy 1979d), these adult floaters probably put more pressure on territory owners than do the subadults. Subadult males, however, are capable of holding territories and breeding (Wright and Wright 1944, Beer and Tibbits 1950, Searcy 1979b) and are known to trespass on territories and to replace missing owners, so subadults must exert some pressure on owners as well.

Considering overall sex ratios and breeding sex ratios in other birds would lead us to expect that the proportion of nonterritorial males is considerably lower in most species than in red-winged blackbirds. Unfortunately, independent estimates of proportions of nonterritorial males are difficult to come by. One exception is a study by Arcese (1987), who was able to count directly floaters in an island population of song sparrows; he found 0.3 floaters per owner in one year, and 0.2 in a second.

A third way of assessing the amount of pressure on territory owners is to examine intrusion rates, the rates at which other males trespass onto territories. Yasukawa (unpublished) estimated mean weekly rates of intrusion at a marsh in Indiana (Figure 3.11) and a prairie in Wisconsin (Figure 3.12). Rates of intrusion were quite variable within a habitat (e.g., 1.3–9.2/hour at the marsh, 0.1–3.0/hour at the prairie) but tended to be highest from the arrival of the females through the early part of the nesting period. In addition, it is clear that rates of intrusion were much higher at the Indiana marsh than at the Wisconsin prairie (Figures 3.11 and 3.12 are plotted with the same Y-axis scale for comparison). Additional data on intrusion rates come from control groups of studies investigating the effects of altering song or epaulets. Table 3.3 presents a sample of such data, together with the mean intrusion rates from the Indiana and Wisconsin studies. The lowest rate is 0.7 intrusions per hour observed on the Wisconsin prairie, and the highest is 6.2 per hour observed by Yasukawa (1981c) on a mixed marsh-upland in New York.

These intrusion rates, especially those from marshes in Indiana and New

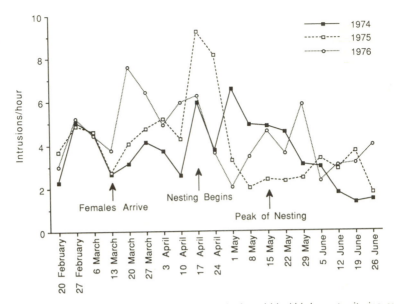

Figure 3.11 Rates of intrusion by other male red-winged blackbirds on territories on a marsh at Yellowwood Lake, Indiana. Observations were made during mornings in 1974–1976. The data shown for each year are means for 22, 23, and 28 territories, respectively. Intrusion rates are high throughout the breeding season but are especially high at about the time nesting begins.

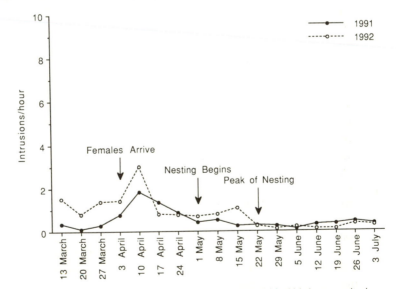

Figure 3.12 Rates of intrusion by other male red-winged blackbirds on territories on an upland site at Newark Road Prairie, Wisconsin. Observations were made during mornings in 1991–1992. The data shown for each year are means for six territories. Intrusion rates are substantiadly lower than on a marsh (see Figure 3.11).

Table 3.3
Rates of Male Intrusion onto Territories in Red-winged Blackbirds.

Intruders/Hour	N	Habitat	Locale	Reference
0.9	10	marsh	Pennsylvania	Peek 1972
1.0	10	upland	New York	Smith 1979
6.2	32	mixed	New York	Yasukawa 1981c
4.0	73	marsh	Indiana	Yasukawa unpublished
0.7	12	upland	Wisconsin	Yasukawa unpublished

York, are high relative to most other territorial species. For example, intrusion rates have been estimated as 0.2/hour in Scott's seaside sparrow (McDonald 1989) and 0.3/hour in pied flycatchers (Searcy et al. 1991, unpublished data). Rates in the redwing range have, however, been found in some other species, such as song sparrows (2.4 intrusions/hour; Arcese 1987).

The high rates of intrusion found in red-winged blackbirds are compatible with the high exclusivity of the territory claimed earlier because territory owners repel intruders so quickly. Observations of 16 separate intrusions showed that territory owners flew at intruders within 18 seconds of an intruder's arrival on average; in all cases the owners were quickly successful in chasing the intruder off (Hansen and Rohwer 1986).

As a final measure of the intensity of competition for territory, we can consider the frequency of escalated fighting for territories. Unfortunately, we can find no quantitative data on rates of fighting, perhaps because rates are low enough to discourage data collection and analysis. Instances of fighting have been described in several studies, and we ourselves have observed many fights over the years, but our impression, based on thousands of hours of observation, is that fighting in red-winged blackbirds is not especially frequent compared to other territorial species. In fact, considering the high proportion of the male population lacking territories, and the frequency of intrusion, fighting seems surprisingly rare in redwings.

The relative infrequency of fighting among male red-winged blackbirds may be related to their possession of dangerous weapons, in particular their sharp bills. The shape of the bill probably evolved as an adaptation for a mode of foraging in which the bill is inserted into vegetation and then gaped open to expose insects (Orians 1985). The male's claws are also formidable, being strong and sharp enough to pierce human skin, as both authors can attest. As evidence for the effectiveness of these weapons, Rohwer (1982) describes the effects of attacks of territory owners on taxidermic mounts of conspecific males: "Flight hits sometimes tore the entire back off a mount; powerful jabbing blows often chipped the glass eyes out of their sockets of hardened clay; similar jabbing blows at the back of the skull would chip away the head and its hardened clay filling; finally, attacking males often pounded into the back of these mounts in the area of the lungs where birds largely lack skeletal shields." Such aggressive responses are the basis for the operation of the "decoy trap" (Smith 1976), which uses a live male redwing as its "bait." As Rohwer (1982) points out, his observations indicate that fighting between two males may be quite risky, even for the probable winner, so males may avoid fighting despite the intensity of competition for territories.

Interspecific Territoriality

Interspecific territoriality refers to the defense of mutually exclusive areas against individuals of other species. Red-winged blackbirds are interspecifically territorial in certain parts of their range, and in those areas this interspecific competition further increases the difficulty of obtaining and keeping a suitable territory. One particular competitor is the closely related tricolored blackbird. Tricolored blackbirds have a much narrower range than redwings, being confined to lowland California and small areas of Baja California and Oregon. In appearance, the only obvious difference between the males of the two species is that tricolors have a narrow white stripe at the bottom of the epaulet instead of a yellow stripe or no stripe as in male redwings. Females of the two species are even more similar in appearance. Male tricolors are slightly smaller than male redwings, about 3% smaller in wing length and about 4% smaller in mass (Orians 1961). Tricolors, like western populations of redwings, nest almost exclusively in marshes. One difference between the two species is that tricolored blackbirds nest in much denser breeding aggregations, often of immense size (Orians 1961).

Territorial interactions between tricolors and redwings were studied by Orians and Collier (1963) at several sites in California. They found that redwings establish their territories earlier in the spring than do tricolors. Later, when the tricolored blackbirds move into a marsh to begin breeding, they are attacked by the male redwings holding territories there. The attacks of the redwings are vigorous, whereas the tricolors show no aggression in return. Individual male redwings are always dominant to individual tricolors. Nevertheless, the redwings fail to exclude the competing species, presumably because the numbers of tricolors per redwing territory are so high. Gradually aggression from the redwings subsides, and the redwing territories are displaced to the periphery of the tricolor colony. Thus, tricolored blackbirds are victorious, apparently, by virtue of their higher densities and persistence.

A second species with which redwings compete for territory is the yellow-headed blackbird. Yellow-headed blackbirds breed over a large part of the interior of North America west of the Mississippi, thus overlapping a substantial part of the range of red-winged blackbirds. Yellowheads breed in marshes, favoring the same types of vegetation as redwings. In this case, the social system of the competitor is very similar to that of the redwing: male yellowheads defend territories of about the same size as do redwings, and they have harem sizes comparable to those found in redwings. In appearance, neither sex of yellowhead resembles redwings at all closely. Male yellowheads have the yellow head that their name implies, and this together with their substantially larger size (they are 17% larger in wing length) and white wing patches make them unmistakable.

As in their competition with tricolored blackbirds, redwings get the jump on yellowheads by establishing territories earlier in the spring, but to no avail. Typically, redwings occupy all the emergent vegetation in a marsh and then are displaced by the arriving yellowheads from those areas with deeper water and sparser vegetation (Orians and Willson 1964). This produces a pattern in which yellowheads occupy the center of a marsh and redwings the periphery. It cannot be population density that decides the outcome of this interaction, as densities of the two species are comparable; instead, it is undoubtedly the greater size of male yellowheads that gives them the advantage. Individual male yellowheads dominate male redwings both off the breeding grounds and in the centers of marshes, but redwings are able to maintain dominance in the periphery, perhaps because this habitat is more valuable to them than it is to yellowheads (Orians and Willson 1964).

In parts of western North America, then, the area occupied by the territories of red-winged blackbirds is constricted by competition from tricolored and yellow-headed blackbirds. Even within the geographical distribution of these latter two species, however, redwings occupy a wider range of marshes than do their competitors and so occupy many marshes exclusively. In eastern North America, redwings do not face such strong interspecific competition from any species. Male redwings of eastern populations sometimes attack common grackles nesting in the same marshes, but the grackles show little aggression in return and do not exclude redwings from any part of the marsh (Wiens 1965). At any rate, most marshes are not inhabited by grackles at all. Redwings are often quite aggressive toward marsh wrens in both the East and West, but we consider this aggression to be antipredator defense rather than interspecific territoriality. Aggression against grackles might also represent antipredator defense, as grackles are known to remove both redwing eggs and nestlings.

Site Fidelity

Most male red-winged blackbirds that hold territory in two successive years are site faithful to the extent that the second year's territory at least overlaps the first. For example, Searcy (1979b) reported on a three-year study in which he observed 39 instances of males holding territories in successive years, and in 34 of these (87%) the returning males were site faithful by the above criterion. Similarly, in a five-year study, Picman (1987) recorded 34 instances of males holding territories in successive years, with 32 cases (94%) of site fidelity. In an eight-year study, Beletsky and Orians (1987a) found males were site faithful in 251 of 281 cases (89%).

All these studies may overestimate site fidelity to some extent, in that site-faithful males are easy to find, whereas males that move between territories may move off the study area, so that they are not counted in the second year. Beletsky and Orians (1987a) provide the best data with which to assess this problem. These authors studied a linear series of marshes approximately 2800 meters from end to end. Most between-year moves (73%) that they observed were less than 200 meters, though there was ample opportunity to observe much longer moves. In one year, a systematic search was made of surrounding marshes, out to a distance of 3500 to 4500 meters from the center of the main study site. Although many males were found who had been banded in the study, mainly at large central feeding traps, no males were located who had ever held territory on the main study site. Thus it seems probable that Beletsky and Orians, at least, overestimated site fidelity very little.

Why are male red-winged blackbirds so often site faithful? We believe the answer lies in "resident's advantage." Resident's advantage refers to the advantage a resident may have in aggressive encounters just by virtue of being a resident. Resident's advantage manifests itself as the ability of a territory owner when on his territory to defeat other individuals that he would not be able to defeat elsewhere. Evidence for a resident's advantage in red-winged blackbirds is the ability of territory owners to repel such a high proportion of intruders, as in Hansen and Rohwer's (1986) study, in which owners defeated intruders on 16 of 16 occasions. Resident's advantage lasting between breeding seasons is implied by the ability of male red-wings to reclaim their former territories in the succeeding year on so high a proportion of opportunities.

Just how high a proportion of returning males succeed in reclaiming their previous year's territory is difficult to estimate because of the problem of finding and counting males who hold territory in one year, survive to the next year, but neither reclaim their old territory nor get a new one. Again, the best data are from Beletsky and Orians (1987a). These authors recorded 251 instances of males returning to reclaim the same territory in the succeeding year, 30 instances of males returning and obtaining a different territory, 39 instances of territorial males who survived to at least one subsequent breeding season but never again obtained a territory, and 11 instances of males territorial one year who failed to get any territory the next year but did manage to claim a territory in a later year. Here the number of surviving males failing ever to get another territory may be underestimated, but the underestimate must be slight as these data give an overall return rate of 60%, which is actually somewhat higher than the 54% survival rate estimated for North American males in general (Searcy and Yasukawa 1981a). This return rate argues that few surviving males were missed. If we assume that all surviving males "wanted" to reclaim their old territories, then their rate of success was 251 out of 331, or 76%. If we assume that those moving to a new territory chose that option rather than being forced into it, then the success rate of males wanting their old territory was 251 out of 301, or 83%. Considering the numbers of males vying for territory, and especially the numbers failing to get any territory at all (see above), it is clear that under either assumption, owners must retain a considerable advantage into the next breeding season in competing for their old territory.

The population studied by Beletsky and Orians (1987a) is unusual in that territorial males visit their territories at least occasionally throughout the year, which may make it easier for owners to retain site dominance. In other populations, however, the percentage of males returning to regain their previous territories is similar. In the Beletsky and Orians study, males survived and reclaimed their old territories in 251 of 544 cases, or 46%. Searcy (1979a) found males surviving and reclaiming their old territories in 34 of 76 cases (45%), and Picman (1987) in 32 of 76 cases (42%). These data show that old owners retain site dominance between years to about the same extent in these other populations as in that studied by Beletsky and Orians.

We argue, then, that males retain a resident's advantage from one breeding season to the next and that it is to exploit this advantage that they show such strong site fidelity. We will take up the question of why resident's advantage occurs later (chapter 7); now we will consider alternative explanations for site fidelity. One possibility is that males return to their previous territories because these territories are superior to all other available territories. There is something to this explanation, as shown by the fact that those males that do move between years tend to be ones whose original territories were inferior. For example, Beletsky and Orians (1987a) found that males who moved had territories in the first year that produced fewer fledglings on average than the territories of males who did not move. If males were site faithful only if they had a superior territory, however, we would expect about half the returning males to move, that half with below average territories. In fact, the fraction moving is far lower than that. Moreover, some males with clearly inferior territories are faithful to them, as for example, four Washington males who returned to territories on which no females had nested the previous year (Searcy 1979b), and two Indiana males who defended latrines for several seasons and were (understandably) never able to acquire any females (Yasukawa unpublished).

Another possible explanation for site fidelity is that males return to a familiar territory because they have accumulated knowledge about it, which makes the territory more valuable to them than any otherwise equivalent territory. Such knowledge could concern locations of food and good nest sites (to demonstrate to females) and where there are safe sites to hide from predators. Most redwing territories are so small and structurally simple that this argument does not seem terribly plausible, yet knowledge about the territory is one of the better supported explanations for why resident's advantage occurs in general. Again, we will return to this subject later. For now we conclude that though other factors may contribute, retention of resident's advantage seems to be the most important reason for the high incidence of site fidelity.

Obtaining a Territory

If site dominance is so strong, how does a male ever obtain a territory in the first place? The only easy route would seem to be to obtain a territory when an old owner dies, and this method does appear to be the single most common one. Yasukawa (1979) observed 37 instances in which a male obtained for the first time a territory on which he was able to breed. Eleven of these males acquired territories of owners that failed to return from migration, and seven more acquired territories whose owners disappeared suddenly during the spring. Another 11 obtained territories that had been vacated by the nonreturn or disappearance of an experienced owner, occupied briefly by an inexperienced male, and then vacated again by desertion. In total, then, 29 of 37, or 78%, obtained their territories directly or indirectly because of the disappearance and presumed death of the original owner. Picman (1987) obtained similar results. In a sample of 41 males obtaining territories for the first time, 34 were recorded as obtaining all or part of a territory belonging to a male that disappeared. One additional male took the territory of an old owner who moved to another part of the study site. Thus 35 of 41, or 85%, of the first-time owners established themselves on vacant territories. Yasukawa (1979) suggested that successful occupation of a vacant territory might be facilitated by prior experience in the area, but detailed observations and experiments by Shutler and Weatherhead (1991a, unpublished) have not supported this suggestion.

If most males obtain territories when they are vacant, it remains true that a few do succeed in wresting all or part of a territory from an owner in residence. Yasukawa (1979) observed seven instances in which new males "inserted" themselves by taking parts of territories from one or more owners; this represents 19% of his sample of new owners. Picman (1987) found six instances of insertion, 15% of his sample. Yasukawa (1979) observed only one instance of a new male taking an entire territory from an old owner, and Picman (1987) none. Thus, the total defeat of an owner by a new male is quite rare, though other instances have been recorded (Nero 1956b).

The tactics by which displacements are achieved are interesting. It seems that, in general, neither partial nor full displacements are preceded by a single, decisive fight, from which the new owner emerges victorious. Rather, displacements usually occur through a process of attrition, in which a persistent intruder gradually wears down an owner (Nero 1956b, Yasuka- wa 1979). Typically, when the interaction is first observed, the old owner is still dominant throughout his territory. The new male repeatedly enters the territory, flying through it slowly and at times even approaching the owner. The owner flies at the intruder, who avoids him but does not quit the terri- tory. Sometimes the intruder remains in the vegetation and is displaced again and again. At other times, the intruder begins to circle upward over the territory, accompanied by the owner, who attempts always to keep above him. Heights of 100 meters or more may be reached before one or the other of the males breaks off and returns to the territory. This type of persistent intrusion was termed "challenging" by Nero (1956b). Many challengers are eventually defeated, but some succeed in establishing dominance, gradually coming to stand up to and return the attacks of the old owner. These obser- vations of displacement indicate that energy reserves and stamina may be as important or more important to success than is fighting ability.

We know little about how males decide where to attempt insertions or displacements, but a few hints have been obtained. Beletsky (1992) sug- gested that floaters could increase their chances by seeking out areas of social instability, i.e., areas where owners are already experiencing diffi- culties with intruders or other owners. As evidence, he showed that floaters were attracted to territories on which he placed multiple models of male redwings; more floaters were attracted to two models than to one, and even more were attracted to four models. Prospecting males may also assess the vigor of defense by territory owners. In support of this idea, Freeman (1987) showed that new neighbors were more likely to intrude onto territories whose owners attacked a model intruder weakly than onto territories of strong attackers.

Conclusions

Male red-winged blackbirds defend territories averaging about 2000 square meters in size during the breeding season, usually in marshes but also in uplands. Defense is accomplished primarily through vocal and visual dis- plays, which are backed up with chasing and fighting when necessary. The spacing behavior of male redwings meets the most stringent definitions of territoriality, in that aggression and display are spatially restricted, and use of space is almost exclusive between males.

Competition for territories is intense. There are large numbers of floater males that lack territories, perhaps as many as two per territory owner. Rates of intrusion onto territories are high, and vacated territories are claimed very quickly. Surprisingly, fighting over territories is not very frequent. In some areas, pressure on territories is intensified by competition from males of two other species, tricolored and yellow-headed blackbirds.

Surviving territory owners show a high degree of fidelity to their previous year's territories. We believe that the primary advantage of site fidelity is that owners retain a resident's advantage from year to year, so that it is easier for them to claim and defend their old territories than other areas. Most males obtain territories initially by finding an open area, one vacated either by the death of the previous owner or (more rarely) by the owner's having moved to a new territory. Displacements of owners by floaters are rare, and when they do occur, are usually accomplished by a process in which the floater gradually wears out the owner in chases and displays rather than by a decisive fight.

Among many unanswered questions concerning territoriality, some of the most interesting to us concern floaters and their interactions with the territorial system. What strategies do floaters pursue in trying to obtain a territory? Do they range widely, or do they concentrate on just a few territories? How much information about territory quality and availability is obtained by floaters, and by territory owners interested in "trading up" to a better territory? When a vacancy appears, is there aggressive competition among floaters to fill it, or does the first male to arrive always win? Why are owners able to repel most floaters with display alone, and why is fighting correspondingly rare? Of course, one reason that these questions are unanswered is that floaters are notoriously difficult to follow and study; nevertheless, we believe that further research on floater behavior would be rewarding.

4

Female Reproductive Success

Understanding the factors that determine female reproductive success is crucial to understanding the mating system of red-winged blackbirds because these factors, whatever they are, should determine where females settle, and female settlement determines whether social polygyny occurs. Reproductive success of female vertebrates has three main components: longevity, fecundity, and offspring survival (Clutton-Brock 1988). Studies of female red-winged blackbirds have provided a wealth of data on fecundity and on offspring survival up to fledging but relatively little data on offspring survival after fledging or on adult longevity. Data on adult female longevity are difficult to gather because of the relatively long lifespan and low site fidelity of female redwings. Data on offspring survival after fledging are even more difficult to obtain because only a very low proportion of the young of either sex return to their natal areas to breed (Orians and Beletsky 1989). Therefore, we will be forced to discuss female reproductive success mainly in terms of success in producing fledglings within single breeding seasons. We will call this measure seasonal reproductive success.

In this chapter we will first discuss the relative importance of the various components of reproductive success in producing variance in seasonal and lifetime reproductive success in female red-winged blackbirds. We will discuss what is known about longevity and fecundity but will concentrate on the components of offspring survival up to fledging, especially escaping the hazards of starvation, predation, and brood parasitism. Second, we will discuss the various factors that influence seasonal reproductive success, such as territory quality, nest defense, and parental provisioning. Again, the relative importance of these factors in determining female reproductive success has implications for patterns of female choice of mates, and hence for explaining the evolution of polygyny. Finally, we will consider whether polygyny really is advantageous to males, by examining the relationship between male pairing success and reproductive success.

Components of Female Reproductive Success

FECUNDITY

If we define fecundity as the number of eggs laid per year per female one year old or older, then fecundity is a function of clutch size and the number of clutches laid per year. Over a female's lifetime, her total fecundity is also affected by the age of first reproduction. We can assume that there is little variation in age of first reproduction among female red-winged blackbirds, as apparently most or all females first breed at the age of one year. Virtually all one-year-old females collected at the end of the breeding season have brood patches (Holcomb 1974), indicating that they have incubated and therefore have attempted to reproduce. As there is little variation in age of first reproduction, this factor cannot contribute importantly to variation in reproductive success.

Clutch sizes of red-winged blackbirds vary from one to six in North America. Variation within a locale occurs for many reasons, including settlement date (Westneat 1992b), date of egg laying (Caccamise 1978), and age of the female (Crawford 1977a). Clutch size can have a direct effect on the number of fledglings produced. For example, Francis (1975) studied clutch size and success of 211 nests in old field, marsh, and wet meadow habitats. The proportion of nests raising at least one young did not vary significantly among clutch sizes, whereas the proportion of eggs producing fledglings and the number of fledglings produced per nest did. Clutches of three and four eggs produced more fledglings that did clutches of one, two, and five eggs.

Female red-winged blackbirds quite commonly lay more than one clutch per year, but in most cases they do so only if the earlier clutch is lost to predation or some other accident. Dolbeer (1976), working in an old field in Ohio, found that 25 of 132 banded females (19%) renested within the same year, but this is a minimum estimate because some renesting may have occurred outside the study area and therefore escaped observation. Two of these females renested twice, and the rest only once. Of the total 27 renestings, only three (11% of renests, 2.3% of females) followed successful nesting. Picman (1981), working in a salt marsh in British Columbia, found a higher proportion of females renesting (45%), though again, most renesting (86%) followed loss of the previous clutch. Still, 10% of Picman's females succeeded in raising two broods in a season, which is much higher than implied by Dolbeer's (1976) results. In a 10-year study, Orians and Beletsky (1989) found that a mean of 5.9% of females raised two broods to fledging each year. Nero (1956a) observed only three cases (3%) of double brooding in five years of observation. Thus, it is clearly rather rare for female red-winged blackbirds to increase their seasonal reproductive success by raising more than one clutch to fledging.

LONGEVITY

As mentioned above, data on longevity are difficult to gather for female red-winged blackbirds because of their long lifespans and low site fidelity. Data on the recovery of dead birds give an estimate of the annual survival rate of female redwings of 54% (Searcy and Yasukawa 1981a). Assuming that survival is fairly constant through adult life, as is typical for small passerines, a study of a cohort of 100 females would have to extend for eight or nine years to determine longevity for all subjects. It follows that adequate studies of longevity must be fairly long-term.

In British Columbia, Picman (1981) found that 62% of 85 surviving females renested in a second year on the same territory as the previous year, 31% nested in a contiguous territory, and 7% moved farther. In New York, Westneat (1992b) found that 26 of 40 surviving females (65%) returned to the same territory in a second year. Females in these studies, then, showed fairly high site fidelity, but their fidelity is nonetheless substantially lower than found for male redwings. Beletsky and Orians (1991) found even lower site fidelity for females on a Washington State site. Females moved a mean of 150 meters (N = 644) between years, equivalent to several territory diameters. A substantial proportion of females (28%) changed marshes between years. Females on average moved more than twice as far following unsuccessful breeding as following successful breeding. Low site fidelity means that adequate studies of female longevity must cover a fairly large area in order to track most surviving females in successive years.

The one study of female redwings that examines a large enough area for a long enough time to estimate longevity adequately is that of Orians and Beletsky (1989). These authors studied several marshes, containing a total of 70 to 80 territories, for a period of 10 years. Out of a sample of 973 females, 55% bred in only one year, whereas at the other extreme one female bred for 10 years. Lifetime reproductive success (in number of fledglings produced) increased linearly with years of breeding (Figure 4.1). Thus, longevity was a very important component of lifetime success for female red-winged blackbirds.

OFFSPRING SURVIVAL TO FLEDGING

Between egg laying and fledging, red-winged blackbird young are lost in large numbers and to a variety of sources of mortality. Table 4.1 gives some

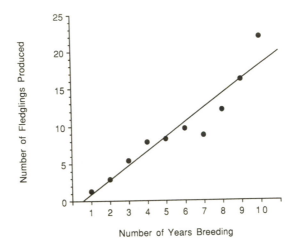

Figure 4.1 Lifetime reproductive success versus number of years breeding for female red-winged blackbirds, from Orians and Beletsky (1989). Lifetime reproductive success (measured as the number of young fledged per female) increases linearly with years breeding (r^2 = 0.90, P < 0.01). Each point represents a mean from several females; the total sample size is 973 females. Data are from a freshwater marsh in Washington.

representative data on such losses, taken from Robertson's (1972) study of redwings nesting in cattail marshes and upland fields in Connecticut. Data for the two types of habitat are given separately. Below we comment on each of the major sources of mortality found by Robertson and compare his findings to those of other studies.

Table 4.1
Sources of Offspring Mortality in Marsh and
Upland Habitat in Connecticut.

Source of Mortality	Marsh	Upland
Total eggs	2418	521
Losses in egg stage		
Failure to hatch	155 (6.4%)	33 (6.3%)
Predation	407 (16.8%)	118 (22.6%)
Desertion	129 (5.3%)	40 (7.7%)
Miscellaneous	108 (4.5%)	35 (6.7%)
Losses in nestling stage		
Predation	236 (9.8%)	100 (19.2%)
Starvation	175 (7.2%)	22 (4.2%)
Miscellaneous	147 (6.1%)	36 (6.9%)
Number fledging	1061 (43.9%)	137 (26.3%)

SOURCE: Robertson (1972).

FAILURE TO HATCH

Robertson found that of those eggs surviving long enough to hatch, 6.4% failed to do so in marshes and 6.3% in uplands. Presumably, these failures to hatch are predominantly due to infertility or embryonic death. Other studies have found slightly higher (Caccamise 1978) or slightly lower (Young 1963) rates of hatching failure. Information on why eggs fail to hatch is provided by Westneat (1992a) for a New York population nesting in marsh and uplands. Westneat found that of 362 eggs that were incubated long enough to hatch, 22 (6.1%) failed to do so. Nineteen of these were broken open and examined; eight had embryos that had died before hatching, seven were dented (perhaps during laying), and four were apparently infertile. Thus, the total rate of loss due to either male or female infertility was on the order of four out of 362, or 1%.

PREDATION

Robertson found predation in both marshes and uplands to be the greatest single source of mortality during both egg and nestling stages. Overall, 26.6% of the eggs and young were lost to predation in the marshes and 41.8% in the uplands. Another way to look at this is that predation accounted for 47.4% of the total mortality in the marsh habitat and 56.8% in the uplands. Table 4.2 shows the percentage of eggs or clutches lost to predators for seven studies in addition to Robertson's. Loss to predation varied from 27 to 53%, with an overall mean of 41%. The result that predation is the most important mortality source is very general (e.g., Smith 1943, Orians 1961, Young 1963, Holm 1973, Caccamise 1976, Weatherhead and Robertson 1977a, Searcy 1979a, Picman 1980a, Orians and Beletsky 1989, Westneat 1992b).

A wide variety of reptiles, mammals, and birds have been suspected of depredating redwing nests. The identity of the predators responsible for nest losses can be important, for example in determining whether predator defense is effective. Predators usually work cryptically, so identifying them can be difficult. Robertson (1972) thought that raccoons were the major nest predator on his marshes, based on direct observation of one raccoon pulling down nests and eating the contents and on the fact that many of the depredated nests had been treated in a similar manner. Mink, another moderate-sized mammalian carnivore, have been suspected as a major nest predator at other sites (Knight et al. 1985, Yasukawa 1989).

Beletsky and Orians (1991) attempted to identify nest predators in their Washington study area based on physical evidence and direct observation but found in the great majority of cases they could not do so. In those cases where identification was possible, black-billed magpies and mice, especially harvest mice, were the two most common predators, with snakes a distant third. These authors believed that magpies were probably responsible for

Table 4.2

Percentage of Eggs or Clutches Destroyed by Predators in Various Studies of Red-winged Blackbirds.

Sample Size	% Predation	Habitat	Locale	Reference
1632 eggs	53	marsh	Wisconsin	Young 1963
2418 eggs	27	marsh	Connecticut	Robertson 1972
521 eggs	42	upland	Connecticut	Robertson 1972
419 eggs	38	marsh	Washington	Holm 1973
1045 eggs	27	marsh	New Jersey	Caccamise 1978
243 clutches	53	marsh	Oklahoma	Goddard & Board 1967
381 clutches	44	marsh	Ontario	Weatherhead & Robertson 1977a
179 clutches	45	mixed	New York	Westneat 1992b
667 clutches	48	upland	Wisconsin	Yasukawa unpublished
558 clutches	44	marsh	Washington	Searcy unpublished

many of the unassigned cases of predation and thus were the most important predator overall.

Picman et al. (1988) placed cameras at artificial nests, made to resemble redwing nests, in salt marshes in British Columbia. Nests were baited with quail eggs, and the cameras were set to be triggered when the eggs were touched. Thirty-seven photographs of potential predators were produced by this setup; 36 were of marsh wrens and one was of an American crow. Physical evidence of predation, especially the discovery of punctured eggs and nestlings with puncture wounds in otherwise undamaged nests, indicated that most predation on natural redwing nests was also due to marsh wrens in this population (Picman 1977a). Marsh wrens have been suspected as the most important predator at other sites also (Ritschel 1985).

A second study using automatic cameras placed at artificial redwing nests was carried out in a freshwater marsh and neighboring uplands in Ontario (Picman et al. 1994). A total of 57 visits by potential predators were recorded in uplands and 158 in the marsh (Table 4.3) Nests in the marsh were classified according to the depth of the water beneath them. Mammals were the major predators both in uplands and in all but the deepest parts of the marsh; the mammals involved were (in order of importance) raccoons, long-tailed weasels, and red squirrels. A variety of birds were also photographed at the upland nests, notably gray catbirds and house wrens. In the deep marsh, the only important predator was the marsh wren.

Other studies have suggested blue jays (Smith 1943) and snakes (Brown and Goertz 1978) as predominant predators in other areas.

STARVATION

Robertson (1972) estimated loss to starvation as 7.2% in marshes and 4.2% in uplands. He noted, however, that diarrhea accompanied some of

Table 4.3

Number of Visits to Artificial Redwing Nests as Recorded by Automatic Cameras at a Marsh and Adjacent Uplands in Ontario.

Species	Upland	Marsh with Water Depth (cm)		
		0–20	21–40	41+
Raccoon	17	64	24	2
Long-tailed Weasel	9	19	10	—
Red Squirrel	9	—	—	—
Blue Jay	5	—	—	—
American Crow	2	3	—	—
Gray Catbird	9	—	—	—
Virginia Rail	—	—	1	—
House Wren	6	—	—	—
Marsh Wren	—	—	3	32

SOURCE: Picman et al. (1994).

these deaths, which suggests that disease or parasitism may have been a contributory cause. In general, it has been customary to ascribe all partial losses in the nestling stage to starvation, that is all cases in which some of the nestlings die and some live (unless partial losses are accompanied by physical evidence of predation). Other studies have used low rates of weight gain as indicators of starvation (Patterson 1991), but again disease and parasitism could produce this symptom also. Thus, it is generally true that disease and parasitism may actually be responsible for many deaths ascribed to starvation. There are several facts, however, suggesting that starvation does contribute importantly to partial losses: partial losses are low in years when food is abundant (Strehl and White 1986) and high when food is scarce (Brenner 1966), artificially enlarged broods experience higher rates of loss than do control or reduced broods (Cronmiller and Thompson 1980, 1981, Fiala 1981), broods fed by the male as well as the female suffer lower losses than broods fed only by the female (Muldal et al. 1986, Yasukawa et al. 1990, Patterson 1991), and losses most often involve the smallest (and youngest) nestlings, which lack the body fat typical of heathly nestlings (Christopher M. Rogers and James D. Hengeveld pers. comm.). As one example, in an unpublished study of nestlings from Yasukawa's Wisconsin prairie, Jennifer Freeman and Brenda Levihn used petroleum ether extraction to estimate percent lipid content and found that apparently healthy nestlings contained 8.9% ($\pm$ 0.29% SE) fat, whereas apparently starved ones contained only 5.9% ($\pm$ 0.54%) fat.

If we agree to accept partial losses as representing starvation, then the percentage loss to starvation found by Robertson (1972) is fairly representative; some studies have found higher losses to this source (e.g., Holm 1973, Caccamise 1978, Patterson 1991), whereas others have found lower (Cronmiller and Thompson 1981, Fiala 1981).

DESERTION

Robertson (1972) found that 5.3% of eggs laid in marshes and 7.7% of eggs laid in uplands died through the desertion of the nest by the female. Desertion of the female during incubation must always doom the clutch because males do not incubate. Presumably, desertion during the nestling period is usually fatal also, but Whittingham (1994) showed that males can successfully rear broods on their own after the females are removed, and a few cases are known in which males successfully reared nestlings to fledging after the disappearance of the female (one in Indiana [James L. Blank pers. comm.] and two in New York [David F. Westneat pers. comm.]). Some of the "desertions" may actually be due to the death of the female. Robertson found no desertions in the nestling stage, however, and other studies agree that desertion is far more likely early in the nesting cycle than late (Case and Hewitt 1963, Young 1963). This implies that most desertion is not due to death but rather to the female's choosing to abandon one nesting attempt in favor of another. It is much more likely that this sort of choice would be advantageous early in the nesting cycle, when more time is available to begin a new attempt and the current clutch has a lower reproductive value than it will have later.

Again, Robertson's data are fairly representative with regard to desertion, with some studies showing higher rates of loss to desertion (Case and Hewitt 1963, Caccamise 1976) and some lower (Young 1963).

MISCELLANEOUS

Other miscellaneous causes of mortality removed a total of 10.5% of the marsh and 13.6% of the upland eggs and nestlings in Robertson's (1972) study. Among these causes were failure of the vegetation supporting the nest, breakage of eggs by the observer, and mowing of upland fields. Other studies have recorded mortality sources idiosyncratic to particular sites and years. For example, Picman (1980a) found a large loss of redwing clutches due to drowning in high tides in salt marshes on the coast of British Columbia. We have observed cases of mortality caused by high winds, hail, flooding, and even a watermelon rind thrown by a camper. Perhaps the most unusual example is the massive nest destruction that occurred in central Washington State in 1980 when the eruption of Mount St. Helens covered the area with about five centimeters of ash (Orians and Beletsky 1989).

BROOD PARASITISM

Robertson (1972) does not mention any losses to brood parasitism in his study, though he may have included some such losses in his miscellaneous category. Parasitism of redwing broods by brown-headed cowbirds does occur, and redwings are clearly an "accepter" species in that they do not remove cowbird eggs (Rothstein 1975), but the frequency of parasitism varies widely among populations (Table 4.2). Even nearby sites can differ markedly; Facemire (1980) found 0% parasitism in a small marsh and 52% parasitism in a beaver pond about 15 kilometers distant. Rates can also vary among years at the same site; Yasukawa (unpublished) found that yearly rates of parasitism in a Wisconsin upland ranged from 2 to 19% over a 10-year period.

Cowbird parasitism might reduce fledging success of female red-winged blackbirds in two ways: first if the female cowbird removes a host egg when she lays her own, and second if the nesting cowbird competes with the host nestlings for food. Even where parasitism is common, however, it may have little effect on redwing breeding success. Weatherhead (1989) found 35% of nests parasitized by cowbirds in a Manitoba marsh but concluded that parasitism did not significantly reduce redwing fledging success or the mass of the nestlings, and did not significantly increase host brood reduction. Instead, parasitized and unparasitized nests had virtually identical reproductive success.

In two of three years studied, Ortega and Cruz (1988) also found that parasitized and unparasitized nests did not differ in number of fledglings produced, whereas in the third year parasitized nests actually produced significantly more fledglings. This latter effect may be due to cowbirds preferring host nests that are built in sites safer from predation. When Ortega and Cruz (1988) examined only nests that escaped predation, parasitism significantly reduced the number of fledglings produced in two of three years, with a nonsignificant trend in the third year in the same direction. We conclude that brood parasitism probably does lead to some decrease in red-winged blackbird reproductive success, though the impact may be small, variable, and difficult to measure.

Offspring Survival after Fledging

As we noted earlier, very little is known about offspring survival after fledging because of the low rates of return of young to their natal areas. Again, the best data are from Orians and Beletsky (1989), who have long-term data on an extensive study area. Over a 10-year period, these authors found that only 8% of the male young and 3% of the female young fledged in the study area returned there to breed. Moreover, most of the adults breeding for the first time on the study area were not banded and therefore must have been fledged elsewhere. This implies that most of the surviving offspring produced on the study area also bred elsewhere and that return rates of fledglings seriously underestimate survival. Another implication of low return rates is that it would take mammoth sample sizes to discern patterns in survival with respect to characteristics of the offspring, their parents, or their natal territories, which explains why little work has been attempted along such lines.

Factors Affecting Seasonal Reproductive Success

Below we review factors that affect the seasonal reproductive success of females, concentrating on those factors that might influence the mating system, especially by affecting female choice of settlement area. These factors act principally by affecting one of the components of offspring survival to fledging.

Physical Environment

Variation in reproductive success among females might be caused by variation in the physical environment at a number of levels; in order of decreas-

Table 4.4
Seasonal Reproductive Success (Number of Fledglings Produced per Nest) of
Female Red-winged Blackbirds in Marsh and Upland Habitat.

Habitat	Reproductive Success	N	Reference
Marsh	1.87	91	Beer & Tibbitts 1950
	0.72	518	Young 1963
	0.81	1112	Case & Hewitt 1963
	1.45	138	Wiens 1965
	0.81	243	Goddard & Board 1967
	1.02	186	Snelling 1968
	1.44	738	Robertson 1972
	0.84	374	Holm 1973
	1.19	164	Caccamise 1976
	1.05	111	Crawford 1977a
	1.24	381	Weatherhead & Robertson 1977a
	0.68	314	Picman 1980a
	0.43	546	Ritschel 1985
Upland	0.80	446	Case & Hewitt 1963
	0.84	162	Robertson 1972
	0.91	186	Dolbeer 1976
	1.64	92	Ritschel 1985
	0.86	765	Yasukawa unpublished

ing scope, these levels are geographical area, major habitat type, male territory, and nest site. As females presumably do not choose among broad geographical areas when deciding where to breed, we will for the most part ignore geographic variation in reproductive success and concentrate on the lower three levels.

The two major habitat types in which red-winged blackbirds breed are marshes and uplands. Table 4.4 shows mean reproductive success per nest from a number of studies performed in the two contrasting habitats. Reproductive success in both habitats is quite variable between studies, and it is by no means clear that one habitat is generally superior to the other. The overall, weighted mean number of fledglings per nest for marshes (0.94) is only slightly greater than that for uplands (0.89), and the difference is not statistically significant (Mann-Whitney $U = 32.5$, $n_1 = 13$, $n_2 = 5$, $P > 0.9$). Three studies have compared reproductive success in marshes and uplands within the same geographical area: one (Robertson 1972) found that reproductive success was higher in marshes than adjacent uplands, one (Ritschel 1985) found that success was higher in uplands, and one (Case and Hewitt 1963) found equal success in the two habitats.

There is little evidence that food production on the territory affects seasonal reproductive success. Orians (1980) found that females nesting on two lakes in Washington State that were highly productive in aquatic insects did not have higher success than females nesting on two other lakes producing virtually no insects. Ritschel (1985) enhanced food abundance on four areas of marsh in California by providing food (millet and meal worms) on elevated trays. Nests in the experimental areas did not show increased fledging success relative to nests in control areas. Rates of weight gain were higher in the experimental areas, however, which may have led to higher postfledging survival. Females often obtain much or all of their food off the territory, so an effect of the territory's food supply on breeding success is by no means inevitable.

A nest's proximity to important nest predators may affect female success. Picman (1980a) found that redwing nesting success increased with increasing distance from marsh wren nests in British Columbia marshes. Similarly, Ritschel (1985) found that both fledging success and nestling growth rates increased with increasing distance from marsh wrens in a California marsh. In both these studies, marsh wrens were believed to be the major predator on redwing nests. Whether the same effect occurs in areas where other predators are of primary importance is unknown.

Several studies have shown relationships between physical aspects of the nest site and nesting success, but often the nature of the relationships differs from study to study. For example, Lenington (1980) found that nesting success tended to increase with increasing density of vegetation, whereas Picman (1980b) and Weatherhead and Robertson (1977a) found just the opposite trend, and Ritschel (1985) found no relationship. Similarly, a number of studies have found that nesting success increased with increasing height of the nest (Meanley and Webb 1963, Holcomb and Tweist 1968, Holm 1973, Brown and Goertz 1978), whereas others found the opposite (Goddard and Board 1967, Weatherhead and Robertson 1977a, Ritschel 1985). Perhaps the most consistent relationship is that, within marshes, nesting success tends to increase with increasing depth of water under the nest; this has been found by Goddard and Board (1967), Robertson (1972), Weatherhead and Robertson (1977a), Brown and Goertz (1978), and Lenington (1980).

Picman et al. (1994) have demonstrated experimentally that water depth affects predation, which as we have seen is the most important factor in determining seasonal reproductive success. In one experiment, Picman et al. (1994) placed four lines of artificial redwing nests perpendicular to the edge of a marsh in Ontario, running from the center of the marsh into the adjacent uplands. Predation was significantly higher in the uplands than in the marsh. Nests escaping predation up to the fourth day of the experiment were, on average, in water of twice the depth as depredated ones. In a second experiment, three lines of nests were placed parallel to the marsh edge, so that water depth varied whereas distance from the marsh edge did not. Predation again decreased with increasing depth. In a third experiment, water depth was held constant while distance from the marsh edge was varied; distance had no significant effect on predation, though the trend was toward lower predation in the center of the marsh. Increasing water depth worked to lower predation by excluding mammalian predators, and certain avian ones as well. Predation rate actually increased somewhat in the very deepest areas relative to areas of medium depth, because of the prevalence of marsh wrens in the areas of deepest water.

Proximity of the nest to a prominent perch also affects reproductive success. Such perches are used by redwings, particularly males, while guarding their territories and nests (chapter 2). Yasukawa et al. (1992b) examined unmanipulated nests in their Wisconsin prairie and found that successful nests (those producing fledglings) were significantly closer to prominent perches (6.8 meters) on average than were depredated nests (10.9 meters), and the closest perches were significantly higher on average for successful nests (3.6 meters) than for depredated ones (2.0 meters). Such correlations need not imply a causal relationship, but in this case experimental evidence supports the effect. Yasukawa (unpublished) placed two-meter-high poles five meters from randomly chosen nests in his Wisconsin prairie and then compared predation on those nests with the survival of controls nests, which did not receive poles. Experimental nests survived significantly longer (19 days) on average than did control nests (13 days; $U = 95.5$, $P < 0.05$) (Figure 4.2). All 20 cases of nest failure were attributed to predation, so the effect of the perches must have been through lowering of predation.

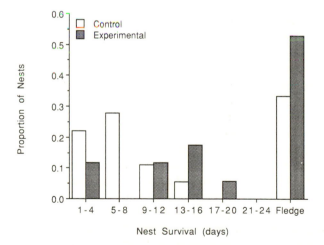

Figure 4.2 Survivorship of 17 experimental nests provided with nearby artificial perches compared to survival of 18 control nests, from Yasukawa (unpublished). Nests near artificial perches survived significantly longer ($U = 95.5$, $P < 0.05$), an effect due presumably to defense by males.

MALE PROVISIONING

As discussed in the previous chapter, there is great variability, both within and among populations, in the extent to which males provision young. Whenever male provisioning does occur, it has significant effects on female reproductive success. In a marsh-nesting population in Indiana, Patterson (1991) found, that pair-fed broods experienced significantly less nestling starvation than did broods fed only by females and produced on average 0.5 to 2.0 more fledglings, depending on the year. Yasukawa et al. (1990) showed a similar effect of male provisioning in an upland-nesting population in Wisconsin: pair-fed broods experienced significantly less starvation (0.46 fewer nestlings starving per brood) and significantly greater fledging success (0.83 extra fledglings per brood) compared to broods fed only by females. Muldal et al. (1986) and Whittingham (1989) also found that male provisioning increased female reproductive success in two other populations where male feeding was common. Finally, Beletsky and Orians (1990) obtained the same result in a Washington population, in which only 6% of males provisioned: nests to which males brought food fledged on average 0.7 more young than other nests on the same territory.

All the above studies were correlational, i.e., the redwings, rather than the researchers, determined which nests would be provisioned. Therefore, it is possible that some factor other than male provisioning, but correlated with it, was responsible for the higher success of male-provisioned nests. For example, it is known that males prefer to provision nests with larger broods (Whittingham 1989, Yasukawa et al. 1990), and so it is possible that male-fed broods produced more fledglings because they were larger to start with rather than because they were fed by the males. Yasukawa et al. (1990) controlled for such problems statistically by performing a stepwise multiple regression analysis, in which number of fledglings produced was the dependent variable, and the independent variables included male feeding rate, female feeding rate, brood size, hatch date, and years of breeding experience in males and females. Male feeding rate was the first variable to enter the regression equation, which argues that male feeding is more important to female reproductive success than the other factors tested.

Female red-winged blackbirds typically do not reduce their own rates of provisioning when males also provision, so that male provisioning is in addition to that provided by females (Muldal et al. 1986, Whittingham 1989, Yasukawa et al. 1990, Patterson 1991). Although pair-fed broods appear to benefit from the increase in feeding visits that result from male provisioning, male-fed broods might also incur a cost in terms of increased conspicuousness to predators, especially as visits to the nest by the larger, more colorful male are likely to be more obvious than visits by the female. Patterson (1991) and Yasukawa (unpublished), however, found that rates of predation were not significantly different for broods provisioned by two rather than one parent.

NEST DEFENSE

As discussed in chapter 2, adult red-winged blackbirds of both sexes help defend nests against predators, through a combination of vigilance, warning, and direct aggression. A number of methods have been used in attempts to test whether nest defense affects reproductive success. The simplest method is correlational. Searcy (1979a) rated the reaction of male redwings to a human observer on a 0 to 7 scale, from no response (0), through various stages of alarm calling, up to repeated direct attacks (7). For 43 males there was no correlation between mean defense score and the proportion of eggs and young taken by predators. Weatherhead (1990a) used a similar scoring procedure and found that nests successful in fledging one or more young had been defended significantly more vigorously than had failed nests. Knight and Temple (1988) measured response of both males and females to crow models placed near nests. Total calls given by the parents correlated significantly with nest survival, but total dives and strikes on the model did not. Yasukawa et al. (1987b) found no correlation between intensity of attacks on a model crow and predation. Thus, the correlational data, taken together, are rather inconclusive.

Yasukawa's (unpublished) experiment showing that artificial perches affect nest survival also provides indirect evidence that nest defense affects breeding success. Provision of perches significantly increased the proportion of time males spent attending the nearest nests. Presumably, this increase in guarding was the major cause of the increase in survival of the experimental nests. Curiously, the evidence that male vigilance reduces predation comes from Yasukawa's prairie study area in Wisconsin, where nocturnal predation by mink and raccoon accounts for the majority of egg and nestling loss due to predators. It is unclear, therefore, how vigilance during daylight hours can reduce predation at night.

Ritschel (1985) placed artificial nests either within one meter of a natural redwing nest or at randomly chosen sites. The artificial nests contained, in one year, a quail egg and a clay egg, and, in a second year, two redwing eggs. In the first year, 86% of 35 randomly placed, control nests were attacked within 14 days, compared to only 51% of 37 nests placed near natural nests. In the second year, 56% of 70 control nests were attacked within two days, compared to only 15% of 39 experimental nests. Both differences were statistically significant. For the experimental nests, time elapsed before the nests were attacked increased significantly with harem size, but this increase did not occur for randomly placed nests. These results indicate that the artificial nests were somehow gaining safety from proximity to redwing

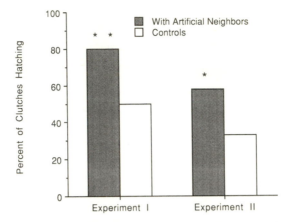

Figure 4.3 Percentage of clutches surviving to hatch for experimental nests that had artificial nests placed nearby, and for control nests with no artificial neighbors; based on data in Ritschel (1985). The experiment was performed in a marsh in California. Each experimental nest had two artificial nests tied into the vegetation within two meters on opposite sides. In Experiment I, each artificial nest contained one quail egg and one clay egg, and sample sizes were 20 experimental nests and 20 controls. In Experiment II, each artificial nest contained two red-winged blackbird eggs, and sample sizes were 21 experimental nests and 21 controls. ** indicates significantly higher survival in experimental nests at the $P < 0.01$ level; * at the $P < 0.05$ level.

nests. The effect may be due partly to active defense by the owners of the natural nests, but it seems also to be due in part to predator dilution. In a separate experiment, Ritschel (1985) placed two artificial nests within two meters on either side of natural nests. In each of two years, eggs in natural nests treated in this way were significantly more likely to survive to hatching than were eggs in control nests (Figure 4.3). As the addition of artificial nests could not have increased the number of defenders, the effect on survival of natural nests must have been due to predator dilution alone.

Picman et al. (1988) performed similar experiments, placing artificial nests containing quail eggs either near (approx. five meters) or far (approx. 20 meters) from a redwing nest. Both the "near" and "far" artificial nests were placed in groups of three, six, or nine (with the redwing nest in the center of each group in the "near" treatment). After four days, 48% of 117 far nests had been depredated, compared to only 20% of 102 near nests ($P < 0.001$). The number of artificial nests in the group had no effect on the probability of predation in either treatment, indicating that predator dilution was not important. Therefore, the increased safety of artificial nests placed near natural redwing nests would seem to be due in this study mainly to defense by the redwings owning the natural nests.

Overall, two conclusions seem justified: (1) Nest defense by the male has some positive effect on female reproductive success; this is supported by

Weatherhead's (1990a) correlational evidence, Yasukawa's (unpublished) artificial perch experiment, and Picman et al.'s (1988) artificial nest experiments. (2) Placement of a nest near other redwings nests increases reproductive success because of predator dilution, active defense, or both; this is supported especially by the artificial nest experiments of Ritschel (1985) and Picman et al. (1988) and also by studies of nesting synchrony (see below).

FAMILIAR NEIGHBORS

Beletsky and Orians (1989b) have proposed that reproductive success in female redwings is enhanced if they nest on a territory whose owner has familiar neighbors. A neighbor is a male defending an adjacent territory, whereas a familiar neighbor is one who was also a neighbor in a previous year. The mechanism suggested is that familiar neighbors cooperate better in nest defense. As evidence, Beletsky and Orians (1989b) showed that females on 71 territories with familiar neighbors produced, on average, 0.2 more offspring per nest than did females on 59 territories without familiar neighbors; however, the trend was not significant. The trend is significant if one excludes the data from one particular marsh, on which reproductive success is low for all females (with or without familiar neighbors). The proposed effect of familiar neighbors is intriguing, but the data thus far must be regarded as only suggestive.

FEMALE AGE

In red-winged blackbirds, older females tend to have redder epaulets and darker chins than one-year-olds (see chapter 9). Using these characteristics to age females, Crawford (1977a) concluded that one-year-old females in Iowa began nesting on average 15 days later than older females and laid significantly smaller clutches. One-year-olds produced less than half as many fledglings per nest compared to older females, 0.73 versus 1.63; this difference was also significant. Orians and Beletsky (1989) classified female redwings as experienced or inexperienced on the basis of whether each had been observed breeding on the study site in a previous year. As noted by Orians and Beletsky, this method of aging is accurate with respect to those classified as experienced, but it will incorrectly label some females as inexperienced, when they have actually bred off the study site previously. Results showed that, though differences in female success were small within years, experienced females did have higher fledging success in each of the 10 years of study.

The lower success of young females may in part be due to their tendency to begin nesting later than older females, as late nesting may be disadvantageous (Dolbeer 1976, Caccamise 1978). An additional factor undoubtedly is that young females are less skilled in providing parental care. Yasukawa et al. (1990) found that the number of feeding visits made per hour increased significantly with years of breeding experience in female redwings. This trend may be due to foraging skill going up with age, as it is known to do in other species of birds (e.g., Orians 1969b, Recher and Recher 1969).

DENSITY AND HAREM SIZE

Using correlational data to analyze effects of density on reproductive success is tricky because expected reproductive success can easily affect density as well as vice versa. For example, in the "ideal free" distribution (Fretwell and Lucas 1969), density within a given patch of habitat is assumed to have a negative effect on reproductive success, yet the correlation between density and reproductive success is predicted to be zero; this lack of correlation occurs because individuals settle more densely in better patches to the extent that success in all patches is equalized. In the "ideal despotic" distribution (Fretwell and Lucas 1969), density has a negative effect on reproductive success, yet the correlation between density and reproductive success is actually positive because resident individuals are able to restrict settlement in better patches below the point at which success is equalized. Thus, correlations between density and reproductive success must be interpreted with caution.

In red-winged blackbirds, distribution models can be applied either to true densities (females per area) or to harem sizes (females per territory). The two measures ought not to be that different, as harem sizes are generally not correlated with territory sizes (Orians 1961, Holm 1973, Yasukawa 1981a), so that large harem size implies high density. Nevertheless, it is possible to get differing results when relating the two measures to reproductive success (Weatherhead and Robertson 1977a). Most analyses have used harem sizes.

Predation losses seem either to decrease with increasing harem size or density (Orians 1980, Picman 1980a) or to show no change (Caccamise 1976, Lenington 1980, Yasukawa et al. 1987b). Rates of brood parasitism from cowbirds are lower on nests placed in the center of a high-density nesting area and are higher on the periphery of such an area or in a habitat supporting a lower density of nests (Friedmann 1963, Robertson and Norman 1976, Linz and Bolin 1982, Freeman et al. 1990). Starvation losses, however, either increase with increasing harem size or density (Lenington 1980, Orians 1980) or show no change (Weatherhead and Robertson 1977a). Most studies have found a positive correlation between harem size and female reproductive success (Haigh's [1968] data in Orians [1972], Goddard and Board's [1967] data in Orians [1972], Ritschel [1985]), though one study found no correlation (Weatherhead and Robertson 1977a). This latter study also found a significant negative correlation between female density (number per area) and reproductive success (Weatherhead and Robertson 1977a). We have found correlations between female density and reproductive success to be either positive or nonsignificant in our Indiana and Washington study sites (Searcy and Yasukawa 1981b).

Thus, not only are the correlational data on harem size and reproductive success difficult to interpret, they are rather inconsistent from study to study. We will examine some experimental data, resulting from manipulations of densities, in the next chapter.

Synchrony

Robertson (1973) found that in each year of his Connecticut study, there was a single temporal peak of nest initiation at each site. He claimed that females nesting at or near these peaks had higher nesting success than females nesting either before or after, but his data were not very convincing. Better evidence that synchrony is advantageous to female redwings was provided by Westneat (1992b) for a New York population. Westneat defined "temporal neighbors" as those females initiating their clutches within two days before or after a focal female. He showed that there was a significant positive correlation between numbers of temporal neighbors and the probability of nesting success, which held true even when nest date was controlled statistically. The effect of synchrony on nest success occurred because synchronous nests suffered less predation. Synchrony did not appear to affect starvation.

Male Reproductive Success and Harem Size

The rationale for concentrating in this chapter on factors affecting female reproductive success is that these are the factors that ought to influence female choice of mate and territory, and female choice in turn ought to determine whether and to what degree polygyny occurs. Factors affecting male reproductive success do not enter into this model because the model assumes that polygyny is always advantageous to males, or in other words that male reproductive success always increases with increasing harem size. If, instead, male reproductive success were to decrease with increasing harem size, males might act to prevent settlement on their territories, and male behavior might then have a major effect on polygyny, if only a negative one. Therefore we need to consider the effect of harem size on male reproductive success.

Most theories of polygyny assume that polygyny is advantageous to males, as it seems logical that males will have higher reproductive success the more females there are working to produce their offspring. This logic is by no means inescapable, however. We can represent male (seasonal) reproductive success (MRS) as:

$$MRS = (FRS)(HS)(WP) + EY \qquad (4.1)$$

where FRS is mean female reproductive success for females within the harem, HS is harem size, WP is within-pair paternity, and EY is the number of young sired by extrapair fertilization. HS appears as a positive term on the right side of this equality, so MRS will increase with HS unless there is a sufficiently strong negative relationship between HS and FRS, WP, or EY.

We have already examined correlations between FRS and HS above. Again, a positive correlation between FRS and HS has been found in three data sets and no relationship in a fourth. This implies that the total number of young raised on a territory (FRS × HS) should increase strongly with HS, and this has been confirmed in several studies (Searcy and Yasukawa 1983, Beletsky and Orians 1987a, Patterson 1991). An example is shown in Figure 4.4 for a population in Washington State.

Only Westneat's (1993a) DNA fingerprinting study of a New York population presents data on WP versus HS. WP was measured as the proportion of the nestlings produced on a territory that were the genetic offspring of the territory owner. Sample sizes were 22 territories in one year and 21 the next. The correlation between WP and HS was slightly positive (Kendall's tau = 0.16) but nonsignificant ($P > 0.25$) the first year and exactly 0 the second (tau = 0.0). This study indicates that within-territory paternity is not negatively correlated with harem size and so cannot reverse the relationship between harem size and male reproductive success. A note of caution is that the range of harem sizes was not great in Westneat's population (maximum = 4); further studies on populations with larger maximum harem sizes would be valuable.

Westneat (1993a) also examined the relationship between EY and HS for his study population. EY was measured as the number of four-to-six-day-old nestlings sired by a male off his territory. EY was positively and signficantly correlated with HS in one year (tau = 0.34, N = 27 males, $P < 0.05$), whereas the correlation was near 0 (tau = 0.06, n = 25, $P > 0.65$) in the second year. The correlation between EY and HS seems to be either positive or zero, so there is no reason to think that production of extrapair young will change the sign of the relationship between harem size and male reproductive success.

Two studies have attempted to measure male reproductive success while accounting for all the terms in Equation 4.1. These are the DNA fingerprinting studies by Gibbs et al. (1990) and Westneat (1993a). Both studies esti-

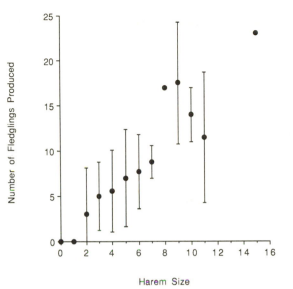

Figure 4.4 Relationship between number of fledglings produced per territory and harem size for one year of a study of a marsh-nesting population in Washington, from Searcy and Yasukawa (1983). Production of young continues to rise throughout the observed range of harem sizes.

mated total numbers of young produced by males within a season as the sum of the number sired and raised on the territory (FRS × HS × WP) and off the territory (EY). For 13 territorial males in their Ontario study site, Gibbs et al. (1990) found the relationship between MRS and HS was slightly positive but nonsignificant (tau = 0.217, P > 0.80). Westneat (1993a) found in one year a positive but nonsignificant relationship for 25 males (tau = 0.13, P > 0.35) and in the second year a positive, significant relationship for 24 males (tau = 0.51, P < 0.001). Note that there is actually rather good agreement between these data sets: in all cases the correlations between harem size and male reproductive success are positive and within a fairly narrow range (tau between 0.13 and 0.51).

We conclude that male seasonal reproductive success does increase with increasing harem size. The evidence is: (1) that the number of females raising young on the male's territory (by definition) increases with harem size, whereas the number of young raised per female, the proportion of those young fertilized by the male, and the number of extrapair young do not decrease with harem size; and (2) that estimates of total numbers of young sired by males within years increase with harem size. The relationship between male seasonal reproductive success and harem size is, however, weaker than one might expect; the looseness of the correlation is a consequence of there being considerable variation in numbers of young raised per female, within-territory paternity, and extrapair fertilizations, all of which variation is largely independent of harem size.

A final consideration is the relationship between seasonal and lifetime reproductive success. Although it is logical to assume that seasonal success can be summed to predict lifetime success, this logic is again not inescapable. The major potential problem is that males that acquire and defend large harems might suffer increased mortality due to greater breeding effort. If harem size were negatively correlated with survival, then harem size might not be positively related to lifetime reproductive success (LRS) despite being positively related to seasonal success.

To determine whether our conclusions on seasonal reproductive success can be extrapolated to lifetime success, we examined LRS data for 50 male red-winged blackbirds from Yasukawa's Indiana marsh. Total number of fledglings produced (our measure of LRS) was positively and significantly correlated both with number of years breeding ($r_s = 0.628$, $P < 0.001$) and with mean harem size ($r_s = 0.414$, $P < 0.01$). This measure of LRS is suspect because it does not take paternity into account. We also found, however, that harem size was positively, rather than negatively, correlated with survival ($r_s = 0.273$, $P < 0.05$). This last result indicates that short-term (seasonal) success in attracting mates does not carry a cost in reduced survival. Therefore, we can extend our previous conclusion that male seasonal success increases with harem size and conclude that success in attracting mates also increases lifetime reproductive success. In other words, male redwings with large harems do what the old Vulcan greeting suggests, they "live long and prosper."

Conclusions

At the start of the chapter, we stated that female reproductive success could be broken into three main components: longevity, fecundity, and offspring survival. Given what is and is not known about female red-winged blackbirds, it is convenient to alter this list to the following: longevity of adults, production of fledglings, and postfledging survival of offspring. The point of this breakdown is that it combines those elements on which we have adequate data (fecundity and survival to fledging) into a single component, the production of fledglings, or what we call seasonal reproductive success. All three of the components undoubtedly have important effects on the lifetime reproductive success, and hence the fitness, of female red-winged blackbirds. In this book, however, our interest in female fitness stems mainly from an interest in how selection has shaped female choice. Female choice of mate and territory may have some effect on longevity, because the female has to live on the territory for some weeks, and on postfledging survival of offspring, because survival probabilities may be affected by the start in life the young get while on the territory, but the main effects are almost certainly on the production of fledglings. Thus, we can probably understand selection on female choice fairly adequately if we can determine the consequences of female choice to seasonal reproductive success.

We have seen evidence that a number of factors influence seasonal reproductive success of female redwings, including most importantly: (1) aspects of the territory and nest site, such as proximity to potential predators and to elevated perches, depth of the water under the nest, and density of the vegetation; (2) the degree to which males provision nestlings; and (3) nest defense by males and neighboring females. In the next chapter we will examine the effects of these and other factors on female choice of territory and mate.

Longevity of adults and survival of offspring after fledging are two subjects about which we obviously know too little. Research on the effects of male and territory traits on these two fitness components would be highly rewarding. At the same time, we feel that there is still a need for more work on factors affecting fledging success, and in particular a need for more experimental work. Too much of the research to date has been correlative, for example correlating vegetation density or male defense effort with fledging success. Results of such correlations tend to vary from study to study, and even when consistent their meaning is ambiguous. It can, of course, be difficult to manipulate the factors thought to affect female reproductive success, but experiments that do so are necessary if we are really to untangle the patterns.

5

Female Choice of Breeding Situation

Female choice of mates has long been a controversial subject. Soon after Darwin (1859, 1871) first proposed female choice as an evolutionary force, his ideas were attacked by Wallace (1889) and others, who doubted that female choice existed. Such doubts persisted well into the twentieth century (Huxley 1938), but in the past two decades they have been buried under an avalanche of data demonstrating that female choice does occur in a great variety of species (Andersson 1994). With the reality of female choice established, the focus of the controversy has shifted to the evolution of female preferences, particularly of female preferences based on male genetic traits. Most participants in this debate can be assigned to one of two schools (Kirkpatrick 1987): the "good genes" school, which holds that female preferences evolve due to selection favoring choice of mates with genotypes that promote viability, and the "Fisherian" school (Andersson 1994), which holds that preferences evolve in response to sexual as well as natural selection, and often favor traits that lower male viability. In simplistic terms, the good-genes argument is that preferences will be favored if they enable a female to obtain a mate who will pass genes for high viability to her offspring, thus increasing the fitness of those offspring. The Fisherian argument states that preferences for male traits that lower viability may evolve because the sons of females with such preferences will inherit the preferred traits and therefore experience high mating success in turn. Population genetics models have been used to support both arguments (for reviews, see Maynard Smith 1991, Andersson 1994).

Female choice in red-winged blackbirds, and in other species with similar social systems, may have a simpler foundation than proposed by either the good-genes or Fisherian hypotheses. If we equate female choice with settlement on the territory of a particular male, then choice has consequences beyond determining which male will sire the female's young. Perhaps most importantly, settlement determines on what territory the young will be raised, and as we have seen, the characteristics of the territory can have important effects on the female's reproductive success. In addition, settlement determines which male will be available to provide parental care for the young, which can again affect the reproductive success of the female. It follows that there will be direct selection on female preferences, that is, selection due to effects on the female's own reproductive success. It is often argued that direct selection on female preferences will outweigh indirect selection caused by effects of preferences on the fitness of the offspring (Searcy 1982, Kirkpatrick and Ryan 1991). As Maynard Smith (1991) states, "in territorial species, and in species with male parental care, female choice is likely to evolve mainly because of the immediate effects of the male's phenotype on her fecundity, and not because of the effect on the genotypes of her offspring."

An important added consideration is that many female red-winged blackbirds copulate with males other than the owners of the territories on which they settle. To some extent, this practice must further lower the relative importance of male genetic quality to the decision on where to settle, in that if a choice based on territory or paternal care leads to a disadvantageous genetic choice, the female can in part escape the genetic cost by copulating with an extrapair male. At the same time, this possibility makes it questionable whether we should continue to equate female settlement with female choice of mate. Presumably, by choice of mate we mean choice of the male (or males) that will sire the offspring. It might therefore seem that choice exercised at the time of copulation should alone be termed female choice. It is clear, however, that this choice is constrained by female settlement, in that the copulating partner is the owner of the territory chosen during settlement in about three-quarters of cases and is a nearby territory owner in the remainder (Gibbs et al. 1990, Westneat 1993a).

Whether or not we term settlement "female choice" is unimportant; what is important is that we keep in mind the sequence of events leading up to breeding. A female first settles on a territory, thereby choosing a social situation, including a territory and a male. The territory provides resources, and the male potentially provides both genes and parental care for the young. Then, when the female is fertile, she chooses one or more males with which to copulate, usually from among the social partner and his nearby territorial neighbors. Choice at this stage affects only which genes will be passed to the offspring. Choice of copulating partner is constrained by the earlier choice made at settlement.

Of the two choices made by the females, it is clearly choice at settlement that determines the form of the mating system in the social sense. Therefore, in this chapter we consider the factors affecting female settlement in detail. First, we consider whether females alone exercise choice at this stage, or whether males also choose. Second, we examine evidence for consistent female preferences for specific territories or males. Third, we consider whether females are influenced in settlement by the number of females already resident on the territory. Finally, we examine the evidence on whether female settlement is affected by such factors as territory quality, quality of male parental care, and male genetic traits.

Which Sex, or Sexes, Exercise Mate Choice?

Darwin (1871) noted as an empirical rule that in most species of animals females exercise mate choice whereas males are much less discriminating: "The exertion of some choice on the part of the female seems almost as general a law as the eagerness of the male." As an explanation for this rule, Darwin could only suggest that male eagerness originated with sessile marine creatures, in which male release of gametes was favored because of the smaller size and greater mobility of sperm compared to eggs. Once adults of some groups acquired mobility, selection favored males that approached females as closely as possible before releasing their sperm. "The males of various lowly-organised animals having thus aboriginally acquired the habit of approaching and seeking the females, the same habit would naturally be transmitted to their more highly developed male descendants" (Darwin 1871). We would today term this explanation "phylogenetic inertia"; male eagerness and female choice is a pattern that is no longer necessarily advantageous but that prevails simply because it has been inherited from ancestors. There may be a good deal of merit to this explanation, but it does not help explain observed exceptions to the general rule (i.e., cases in which males are more discriminating), nor does it predict whether red-winged blackbirds should be among those exceptions.

Building on the work of Bateman (1948), Trivers (1972) advanced an explanation for the predominance of female choice that did attempt to encompass the exceptions. According to Trivers, the factor determining which sex will choose is parental investment, defined as any investment that increases the fitness of one offspring and lowers the parent's ability to invest in other offspring. The sex with greater parental investment per offspring becomes a limiting resource for the sex investing less; the former sex then exercises mate choice, while individuals of the latter sex compete among themselves for opportunities to mate. This parental-investment hypothesis explains the predominance of female choice, as females in most species contribute greater parental investment than males, and also seems to explain why male choice occurs in certain species, such as polyandrous shorebirds, in which males provide more parental care than do females.

To apply Trivers's hypothesis to red-winged blackbirds, we need to decide which sex in this species contributes greater parental investment. As we have seen (chapter 2), female red-winged blackbirds alone provide certain aspects of parental care: they alone build nests, lay and incubate eggs, and brood young. Females also do the majority of the work of feeding the young, which is usually thought to be the most energy-demanding part of parental care. Males, however, probably contribute more to nest defense, both in vigilance and especially in actively attacking predators. The major cost of predator defense may be risk, rather than energy drain, which makes it difficult to balance the cost of defense against the energy costs of other aspects of parental care. It seems likely that female red-winged blackbirds overall provide greater parental investment than do males, but because of the difficulties of adding up investments in different activities, this would be hard to prove.

This difficulty in summing up investments is not peculiar to red-winged blackbirds; as Clutton-Brock and Vincent (1991) remark, "relative parental investment is usually impossible to measure in species where both sexes invest in their offspring." Clutton-Brock and Vincent argue that, at any rate, the mode of sexual selection is more closely determined by the potential rates of reproduction of the two sexes than by relative parental investment. Potential rate of reproduction is the maximum number of offspring that parents can produce per time. The sex with the lower potential rate of reproduction is limiting to the other sex, so the former sex ought to exercise mate choice while the latter shows intrasexual competition. This principle is quite successful in explaining the occurrence of male choice versus female choice in different species (Clutton-Brock and Vincent 1991).

The potential rate of reproduction (per season) of female redwings is easily estimated. As discussed in chapter 4, female red-winged blackbirds are sometimes able to raise two clutches to fledging in one breeding season, though they do so in less than 10% of cases (Dolbeer 1976, Picman 1981, Orians and Beletsky 1989). Two successful broods in one season seem to be the maximum possible. Maximum clutch size is six, which gives 12 fledged young as the outside upper limit for female reproductive success within a breeding season.

Male red-winged blackbirds have been observed with up to 15 females nesting per territory within one season. If each female produced 12 fledglings, then a male with 15 females could potentially have 180 young raised on his territory. Of course, if it is unlikely that a single female is able to raise 12 young in one year, it is even less likely that all members of a harem of 15 would succeed in doing so. The DNA fingerprinting data show that about 75% of the young, on average, are sired by the territory owner (Gibbs et al. 1990, Westneat 1993a), but our maximal male potentially could make up for the loss of genetic parentage on his territory by siring young on other territories.

We conclude that the maximal rate of reproduction in female redwings is on the order of 12 offspring per season, compared to a maximal rate on the order of 180 offspring per season in male redwings. Undoubtedly, neither maximum is ever attained, but maximum attained rates are also greater for males than females. Payne (1979) found in a Michigan population that the maximum number of fledglings produced in one season was 11 for males and four for females. The maxima attained in Yasukawa's Indiana population were 14 for males and four for females. Harem sizes were moderate in these two populations; maximum rates attained by males should be higher in populations with larger harems, and the disparity between male and female rates even greater.

Female red-winged blackbirds, because they have lower maximum rates of reproduction, are predicted to be the more discriminating sex in mate choice. This prediction is in accord with observations of redwing behavior. Females arrive on breeding sites well after males have set up territories, and we have observed them to visit several territories, over a period of several days, before settling. Other evidence for female choice will be considered below.

Even if male redwings are less discriminating than females, it is still possible that they exercise some discrimination in mate choice. Burley's (1981, Burley et al. 1982) work with monogamous rock doves and zebra finches shows that it is possible for both sexes to exercise mate choice in the same species. Male red-winged blackbirds could exercise choice either by aggressively preventing particular females from settling on their territories, or by refusing to copulate with particular females. In the context of the social mating system, it is the former possibility that is of interest.

Male reproductive success seems to increase with increasing harem size (see chapter 4), and therefore aggressive rejection of prospecting females seems unlikely to be advantageous to males. Moreover, aggressive rejection ought to be observable if it occurs. Male aggression toward females can be seen in red-winged blackbirds. For example, territorial males sometimes attack resident females, when the resident females are themselves harassing new females seeking to settle (Nero 1956b, Searcy 1986). Our presentations of stuffed females (Yasukawa and Searcy 1982) occasionally elicited such attack by territorial males. This male behavior is not aggressive rejection of settlers; on the contrary, males seem to be encouraging settlement by inhibiting aggression from resident females. Another context in which males attack females is in "sexual chasing," a behavior marked by "aggressive pursuit by the male and rapid elusive flight by the female" (Nero 1984). A male in hot pursuit may strike or grab the fleeing female. Often, several males join the original chaser. The male beginning the chase is almost always already paired to the female being chased, i.e., she has already settled on his territory (Nero 1984, Westneat 1992a). Thus, sexual chasing does not represent aggressive rejection either. Finally, males sometimes attack females that attempt to forage on their territories, but such aggression seems unrelated to rejection of prospective mates because these females are already nesting on neighboring territories (Nero 1956b).

In conclusion, parental-investment theory predicts that in red-winged blackbirds, females ought to be more discriminating than males in choice of mates, and this prediction is in accord with observations of redwing behavior. The degree to which males also discriminate seems to be low.

Consistency of Female Preferences

One approach to studying female choice of breeding situation is to search for evidence of consistency in female settlement patterns. In this approach, one produces two different, and presumably independent, measures of female preferences for a given set of territories or males and then calculates the correlation between the two measures. If a significant positive correlation is found, this is evidence that the females are exercising consistent preferences, though the analysis may tell us little about the basis of these preferences. A common weakness of the approach is in the assumption that the two measures of preference are independent.

An example of this type of analysis is provided by Garson et al. (1981), who worked with data from Holm's (1973) study of a marsh-nesting population of red-winged blackbirds in eastern Washington State. Garson et al. tested a number of predictions of the polygyny threshold model, among them that "the ordering of the males by harem size must . . . agree with the ordering by first mates, by second mates, etc." (Altmann et al. 1977). Here total harem size is one measure of female preference, and order of choice by females of a certain rank the second. Garson et al. (1981) used first egg dates to estimate date of settlement, thus assuming that first egg date is correlated with date of settlement, and that date of settlement reflects female preferences. Results confirmed the prediction: in most cases, statistically significant negative correlations were found for harem size versus mean first egg dates for primary, secondary, and tertiary females (see Figure 5.1). Orians (1980) found similar relationships in the same population, whereas Lenington (1977, 1980), working in Illinois and New Jersey, did not. It should be noted that though the Garson et al. result is indeed consistent with the polygyny threshold model, it is also consistent with several other polygyny models. The two measures of female preference are not completely independent, however, in that females of a given status rank are also included in the calculation of total harem sizes. Even with random female settlement, a territory that obtained a primary female very early would be likely to accumulate more females than one that obtained a primary female very late, thus producing a negative correlation between settlement date by primary females and harem sizes. It seems probable to us that in this case, the nonindependence of the preference measures would not alone be sufficient to cause the observed correlations, but further analysis would be needed to prove this.

Searcy (1979a) calculated correlations between the numbers of females choosing particular areas of marsh habitat in successive years. The data were from the same region of eastern Washington studied by Holm (1973). Here the number of females choosing given areas in one year and the number

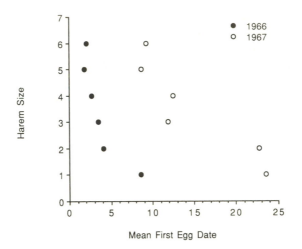

Figure 5.1 Relationships between mean first egg date and harem size for primary females in each of two years, for a marsh population in Washington, from Garson et al. (1981). Both first egg date and harem size can be considered measures of female preference in settlement. A negative correlation is expected if females rank breeding situations in the same order on both measures, and thus a negative correlation is evidence of consistent female preferences. The correlations are negative and significant (1966: $r_s = -0.89$, $P < 0.05$; 1967 $r_s = -0.94$, $P < 0.05$).

choosing the same areas in a second year are treated as independent measures of female preferences. Highly significant, positive correlations were found for two comparisons involving one-year intervals ($r_s = 0.647$ for 1974 versus 1975, N = 42, $P < 0.001$; $r_s = 0.711$ for 1975 versus 1976, N = 43, $P < 0.001$). Aside from the possibility of nonindependence (which we consider below), the positive correlations could be explained by female preferences for specific areas or by female preferences for specific males, as males often hold the same areas in successive years. Two facts argue for the greater importance of area preferences. First, correlations between harem sizes of individual males in successive years were considerably lower than the area correlations and were only statistically significant in one of two comparisons. Second, the correlation in area harem sizes after a two-year interval (1974 versus 1976) was of the same magnitude as for the single-year intervals, despite the fact that less than half as many males would be expected to reclaim the same territories after a two-year interval as after a single-year interval.

Yasukawa (1981a) performed a similar analysis on four years of data from a marsh population of red-winged blackbirds in Indiana. Significant, positive correlations between area harem sizes were found between successive years ($r_s = 0.764$, 0.697, and 0.721 for three one-year intervals; N = 30, $P < 0.01$ in each case). The date of the first nest on a territory was used as another measure of female preference. Correlations between first nest dates for arbitrary territories in successive years were 0.528, 0.477, and 0.499 (N = 30, P < 0.05 in each case). In addition, Yasukawa (1981a) looked at first-nest-date and harem-size correlations for both of the two-year intervals and for the one three-year interval that his data allowed. Contrary to what Searcy (1979a) found in Washington, the harem-size correlations tended to decrease as the time interval increased, and the same was true of the first-nest-date correlations. These trends are consistent with a greater effect of male traits on female choice in Indiana; that is, the correlations may decrease because fewer and fewer males are returning to the same sites.

Picman (1981) examined the correlation between numbers of females settling on actual male territories in two successive years using data from a salt-marsh population in British Columbia. Picman restricted his analysis to just those territories that were not owned by the same males in both years, in this way eliminating any effect of male attractiveness on the correlation. The result was a significant, positive correlation (r = 0.58, N = 16, P < 0.01). This result again argues that the between-year correlations in female preferences are due more to female choice of territories than to female choice of males.

We now turn to the problem of nonindependence in the year-to-year correlations of harem sizes. An important source of nonindependence in this type of analysis arises from female site fidelity. If females are faithful to breeding sites regardless of territory quality, then areas in which many females breed one year will tend to have many in the next, even without any female choice based on traits of territories or males. As it is not obvious whether this hypothesis alone can account for the observed correlations, we undertook computer simulations to address this question (Yasukawa and Searcy 1986). Our first, most basic model assumed site fidelity but no other female preferences, whereas subsequent models made more elaborate assumptions about female choice.

All our models had certain features in common. First, females were assumed to occupy a series of territories arranged in a circle, so that each territory had just two neighboring territories, one on each side. This arrangement was nearly universal at both our Washington (Searcy 1979a) and Indiana (Yasukawa 1981a) study sites because males in both areas were defending territories in belts of emergent vegetation along the margins of lakes. Second, all models were initiated with one of two observed distributions of arbitrary territory harem sizes, one from Washington (42 territories) and one from Indiana (30 territories). Third, a run of a model simulated the events occurring between one breeding season and the next. These events began with a certain fraction (M) of the females being randomly chosen to die; these females were replaced with an equal number of first-time breeders. Next, a fraction (F) of the surviving females were randomly chosen to be

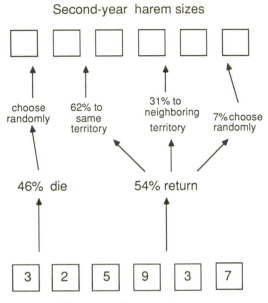

Figure 5.2 Flow chart of the "random choice" model from Yasukawa and Searcy (1986). The model starts with a certain distribution of first-year harem sizes, and generates a distribution of harem sizes on the same territories in the succeeding year. A percentage of the females in the initial distribution (46% in this example) are randomly chosen to die between years. The remaining females survive between years. Of those surviving, a percentage is chosen to be site faithful, i.e. to return to their previous year's territory (62% in this example). Another percentage is chosen to be semi-site faithful, i.e., to return to a territory adjacent to their previous territory (31% in this example). Those dying between years are replaced by an equal number of first-time breeders; these new breeders and the non-faithful survivors choose second-year territories randomly. We then calculate the correlation between the first- and second-year distributions.

"site faithful," that is to return to the same territory they occupied the previous year. Another fraction (SF) of the survivors were chosen to be "semi-faithful," meaning that they settled on a territory adjacent to their previous territory. Finally, the first-time breeders and nonfaithful survivors were settled on territories, according to assumptions that differed between models. These steps generated a second-year distribution of harem sizes, which could then be correlated with the first-year distribution.

Our first model is outlined in Figure 5.2. What distinguishes this "random choice" model is that first-time breeders and nonfaithful survivors both choose territories randomly; hence, apart from site fidelity, the model assumes no female preferences. Initially, values for F and SF were taken from the best available study of return rates of color-banded females, that done by Picman (1981) in British Columbia. Picman (1981) found 62% of surviving females to be site faithful and an additional 31% to be semifaithful. Similarly, the initial value of M was based on the best available estimate of female mortality, calculated from the U.S. Fish and Wildlife Service's records of the recovery of dead birds (Searcy and Yasukawa 1981a); this analysis estimated annual mortality of female redwings to be 46%. With these assumptions, the model was run for 100 consecutive years and gave mean between-year harem-size correlations of 0.322 (SE = 0.006) for Indiana and of 0.362 (SE = 0.005) for Washington, both well below the observed range of correlations. We also ran versions of the model in which we systematically varied S (site fidelity) and M (mortality) and found that even with unrealistically high values of site fidelity and unrealistically low values of mortality, the correlations produced were lower than those observed (Yasukawa and Searcy 1986). We conclude that site fidelity alone is not sufficient to account for the between-year correlations in harem sizes.

Our second model added the assumption that some classes of females show consistent preferences based on characteristics of territories and/or males. This model is outlined in Figure 5.3. As in the previous model, most

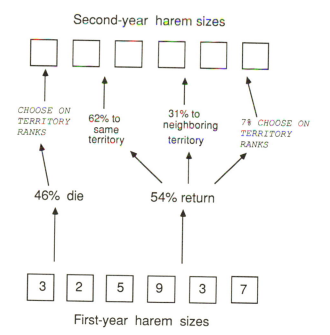

Figure 5.3 A flow chart of the "directed choice" model, from Yasukawa and Searcy (1986). The model is identical to the previous, random choice model (Figure 5.2) except that first-time breeders and non-site faithful surviving females are assumed to choose territories based on a ranking of territory quality. Females choosing on territory ranks are assumed to sample N territories adjacent to a randomly selected point and to choose the highest ranking one subject to certain constraints (see text).

surviving females were either site faithful or semifaithful. What is new in the model is that nonfaithful survivors and first-time breeders chose territories based on a shared ranking rather than settling randomly. In practice, the ranking of territories was based on the original, observed distributions of harem sizes, i.e., we assumed that the territory with the largest harem was the highest-ranked territory, the territory with the second-largest harem was the next highest ranked, etc. Individual females were allowed to sample the N territories adjacent to a randomly selected point, with N varying between two and 20. To allow for the possibility that females might make errors in evaluating territories, we defined an additional parameter, E, the maximum difference in suitability that females are unable to detect. Therefore, if the best territory a female encountered had rank R, she could not discriminate between that territory and any other with rank R − E or better. If a female encountered two or more territories with ranks between R and R − E, she was assumed to choose randomly among those territories.

Model 2 was run with our original, empirically derived estimates of S, SF, and M. As expected, the year-to-year correlations in harem sizes produced by this model were larger than for our first model, showing the effects of assuming consistent female preferences. As also seems reasonable, correlations tended to decrease as error (E) increased, that is, as females became worse at discerning differences between territories (Figures 5.4 and 5.5). Correlations tended to increase as the number of territories sampled, N, increased, but only for Ns up to a limit of about six in Washington (Figure 5.4) and eight in Indiana (Figure 5.5). With optimal parameter values, the Washington simulations produced correlations in the observed range (0.680 from the model versus 0.647 and 0.712 observed). Even with optimal values, however, the Indiana simulations produced correlations somewhat lower than those observed (0.603 from model versus 0.697, 0.721, and 0.764 observed). Additional simulations demonstrated that correlations could be increased further by making the territory ranking more fine grained and thus breaking the many ties that existed in the original ranking. Even with this alteration, the Indiana simulations produced correlations slightly below the observed.

In addition to using about 95% of the computer time available at Beloit College, our models clearly demonstrated that site fidelity alone is not sufficient to account for year-to-year correlations in harem sizes. At the same time, the results are compatible with the existence of consistent female preferences based on male or territory characteristics. There may be mechanisms other than such preferences, however, that would also explain the year-to-year correlations. One such mechanism is "female copying," a behavior in which certain females copy the choice made by other females (Pruett-Jones 1992). Female copying has been demonstrated to occur in other species of birds (Gibson et al. 1991) and in fish (Dugatkin 1992). In redwings, female

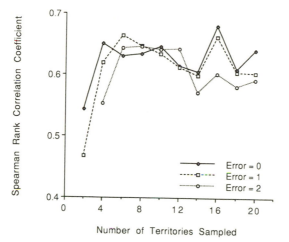

Figure 5.4 Correlations between harem sizes on arbitrary territories in successive years generated by the "directed choice" model, starting with the distribution of harem sizes from Washington State (Yasukawa and Searcy 1986). "Error" is the difference in territory ranks that females are unable to detect, i.e., females choose randomly between territories whose ranks differ by this much or less. When females sample a moderate number of territories (>4), the model generates between-year correlations in the range actually observed in Washington.

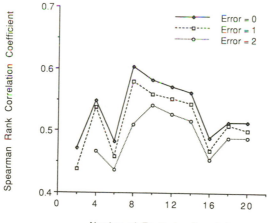

Figure 5.5 Correlations between harem sizes on arbitrary territories in successive years generated by the "directed choice" model, starting with the distribution of harem sizes from Indiana (Yasukawa and Searcy 1986). "Error" is the difference in territory ranks that females are unable to detect, i.e., females choose randomly between territories whose ranks differ by this much or less. With optimal parameter values (females sample eight territories with no error), the model generates between-year correlations almost as high as are actually observed in Indiana.

copying could work either by late breeders copying the choices of early breeders in the same season, or by females in one breeding season copying choices of females in the previous season, assessing these choices by the positions of old nests, which often survive on territories from one year to the next (Erckmann et al. 1990). Of course, female copying is a type of female preference, with many of the same implications for the evolution of mating systems as preferences based directly on territory or male characteristics. Nevertheless, the distinction between female copying and direct preferences is an interesting one.

Female Site Fidelity

Our random-choice model assumes that females breeding for the first time and returning females who are not site faithful choose territories randomly. Note that even with these assumptions, some element of female choice is possible, in that returning females could still choose whether or not to be site faithful. We might expect females to decide whether to be site faithful based on factors that affect their reproductive success (Beletsky and Orians 1991). Decisions about site fidelity might in turn affect the social mating system.

Beletsky and Orians (1991) investigated female site fidelity using 12 years of data from their Washington study sites. Their measure of between-year site fidelity (or lack thereof) was the distance between a female's last nest in year x and her first nest in year x + 1, which we will call "distance moved." Distance moved was substantially greater for females if their last nest in year x was unsuccessful (mean = 237 meters, N = 260) than if it was successful (mean = 105 meters, N = 384, P < 0.001). Females were also more likely to mate with a new male, and to move between marshes, when their last nest in year x was unsuccessful. As long as there is some correlation in site quality from year to year, this pattern of moving farther following nest failure should improve female reproductive success. Year-to-year correlations in the probability of nest success on specific sites have indeed been demonstrated (Searcy 1979a). This pattern of between-year movement should also produce a net tendency for females to vacate poor areas and accumulate in good ones, thus increasing the degree of polygyny.

Effects of Resident Females on Settlement

The presence of females already resident on the territory might either encourage or discourage further settlement. A priori arguments can be made for either alternative. On the side of a positive effect, there is the possibility of female copying, mentioned above, in which certain females might use the settlement pattern of other, more experienced females as a cue to which territories or males are best. Another possibility is that there is a net benefit to polygyny, most likely because grouping by females depresses predation rates. If polygyny has a net benefit, we would expect females to seek out and settle on territories already chosen by others. On the side of a negative effect, there is the possibility that polygyny has a net cost, because the costs of sharing male parental care and the resources of the territory outweigh the benefits of mutual defense. If polygyny has a net cost, we would expect females to avoid settling where other females have already settled. Avoidance of already-settled areas by new settlers produces what can be called "passive density limitation" (Searcy 1988a). Another possibility is "aggressive density limitation," in which aggressive behavior by resident females deters further settlement by other females. We might expect females at least to attempt aggressive limitation if settlement by new females imposes a net cost on those already resident.

Behavioral observations of female redwings suggest that aggressive limitation may indeed be attempted. Resident females are often seen to attack females that are seeking to settle on their mates' territories (Nero 1956a, b; Lenington 1980; Searcy 1986). They will also attack taxidermic mounts of female redwings placed on their mates' territories (LaPrade and Graves 1982, Yasukawa and Searcy 1982), and they will approach and display aggressively in response to playback of female vocalizations (Beletsky 1983a, b). Of course, female aggression may have functions other than density limitation, such as defense of subterritories within the males' territories or establishment of dominance relationships (see chapter 9).

We made an early attempt to determine whether females influence each other's settlement by examining the distribution of time intervals between nesting attempts within and between territories (Yasukawa and Searcy 1981). We reasoned that if nesting on the same territory with other females has a net cost, then increasing the temporal spacing between nesting attempts within a territory would be one way for females to minimize the cost. Conversely, if nesting with other females has a net benefit, then nesting close in time would be a way of maximizing the benefit. We used the temporal distribution of nesting attempts on different territories as our control, under the assumption that females on the territories of different males within the same locality would not directly affect each other's timing but would be subject to the same environmental influences. We performed the analysis on three years of nesting data from Yasukawa's Indiana site and on two years of data from Searcy's Washington site.

Figure 5.6 illustrates the results for Indiana only, as cumulative frequency distributions of time intervals between first egg dates for all pairs of nests in a given year and locality. There is a tendency for short time intervals to be

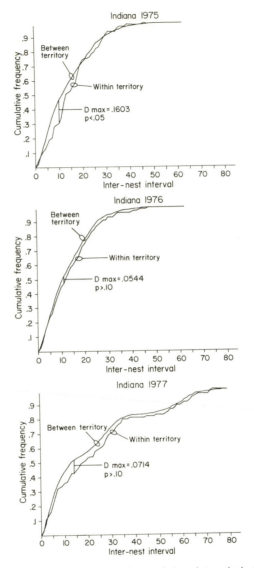

Figure 5.6 Cumulative frequency distributions of time intervals between nest-starts within and between territories in Indiana, from Yasukawa and Searcy (1981). If females avoid settling close in time to other females on the same territory, short within-territory intervals should be rare relative to short between-territory intervals. There is some evidence of the predicted pattern in these data, although only in one year (1975) is the trend significant ($P < 0.05$ according to a Kolmogorov-Smirnov d_{max} test). For two years of data from Washington (not illustrated), there was no difference in the distribution of intervals within territories and the distribution between territories according to Kolmogorov-Smirnov d_{max} tests.

rarer between nests started on the same territories than between nests started on different territories in all three years, but the difference in cumulative frequency distributions was significant ($P < 0.05$) in only one year. For the Washington data, in both years the distribution of intervals between pairs of nests started within the same territory is very nearly congruent with the distribution of intervals between nests started on different territories. The results, then, indicate that females in Washington do not affect each other's timing at all, whereas there may be a weak, negative effect in Indiana.

The first experimental test of whether female red-winged blackbirds influence each other's settlement was performed by Hurly and Robertson (1985) using a site in Ontario. These authors mapped out 13 male territories and randomly chose seven as experimentals and six as controls. From the experimental territories, they removed resident females by shooting on five days: 23 and 25 May, and 6, 9, and 18 June. A few females proved too elusive to be shot, but the great majority of residents were removed, a mean of 2.7 per territory. Females on control territories were left undisturbed. The number of new nest initiations was used to indicate the number of new females settling on territories. The control territories had a mean of 0.5 new nests versus a mean of 1.6 for experimental territories ($P < 0.01$). This result is compatible with aggressive density limitation, i.e., resident females normally deter further settlement through aggression, and this deterrent is removed when the residents are removed. The result is also compatible with passive density limitation, with new settlers normally avoiding areas already occupied, and this avoidance disappearing when the residents are removed. There is, however, an alternative explanation that does not assume that females directly affect each other's settlement, namely that the male territory owners increase rates of advertisement when their resident females are removed, and that increased display attracts additional settlement (see below).

Searcy (1988a, unpublished) has carried out a series of similar experiments, with dissimilar results. All these experiments were performed on marsh populations in Pennsylvania. In his first experiment, Searcy (1988a) captured the first-settling female on each territory in a marsh, and at the time of capture he randomly determined whether the female would be removed (producing an experimental territory, $N = 8$) or immediately released (producing a control territory, $N = 10$). These females were captured in the period after settlement but before nesting.

Removed females, in this and subsequent experiments, were held captive and then released at the end of the breeding season. To check the possibility that removal of females increases male display rates, observations of song rates were made over a nine-day period after the end of removals. Owners of experimental territories sang 27 ± 7 songs (mean $\pm$ SD) compared to only 15 ± 7 for owners of control territories. This increase in song rates was statistically significant (Mann-Whitney $U = 8$, $P < 0.005$). Despite the increase in male singing rates on experimental territories, removal of primary females had no effect on subsequent female settlement. A mean of 1.5 females settled on experimental territories after the primary female, compared to 1.8 on control territories (Figure 5.7a). This difference runs in the direction supporting a positive effect of residents on subsequent settlement, but the trend is not significant (U = 36, P > 0.10). Remaining harem sizes, the number of females remaining after the removals, averaged 2.8 for control territories and 1.5 for experimental; this difference was significant (U = 14, P < 0.02). Removal of the primary female also had no effect on the timing of nesting by the next female; the median first egg date of secondary females was 17 May for both experimental and control territories.

In a second experiment, Searcy (1988a) attempted to follow more closely the design of Hurly and Robertson (1985) by removing females later in the season and removing multiple females per territory. Twelve territories were chosen on a second marsh in the same general area, and half were randomly designated as experimentals and half as controls. Removals began when the earliest nesting female was in the late incubation stage (7 May) and continued until 19 May, when all the original residents had been removed. A mean of 1.7 females were removed per experimental territory. After removals, an identical number of new females settled on experimental and control territories, a mean of 0.7 per territory for both (Figure 5.7b). The mean remaining harem size for experimental territories (0.7) was significantly lower than the mean for control territories (2.3, U = 4, P < 0.02). Thus, there was again no evidence that resident females had any effect on subsequent settlement.

In a third experiment, later-settling females were removed, whereas the primary female was allowed to remain (Searcy unpublished). Fourteen territories were chosen for this experiment, with seven randomly designated as controls and seven as experimentals. At least one secondary female was removed from each experimental territory, and 1.6 females were removed on average. The mean number of females settling after the first capture on a territory was higher for experimental territories (0.6) than for controls (0.3), but the difference was not significant (U = 16, P > 0.10) (Figure 5.7c). Remaining harem sizes were significantly smaller on experimental territories (mean = 1.7) than on controls (mean = 2.7, U = 7.5, P < 0.01). This third experiment, then, agreed with the first two in indicating no effect of resident females on settlement.

Other studies have been designed specifically to look for aggressive limitation. Searcy (1988a) measured the aggressiveness shown by first-settling females toward a taxidermic mount of a female redwing placed near the nest. If aggressive limitation occurs, we would expect a negative correlation between aggressiveness of the first-settling female and harem size. Instead, a positive correlation was found, which, however, was not significant. In a separate experiment, Searcy (1988a) implanted first-settling females with testosterone to increase their aggressiveness. Treatments were in the form of

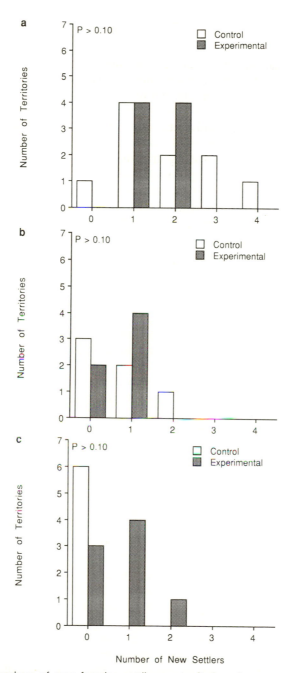

Figure 5.7 Numbers of new females settling on territories after some or all of the resident females had been removed from territories on Pennsylvania marshes, from Searcy (1988a, unpublished). In the first experiment (a), the primary female only was removed from experimental territories; in the second (b), a mean of 1.7 early-settling females were removed; and in the third (c), a mean of 1.6 late-settling females were removed. In none of the experiments was there significantly more settlement on experimental territories than on controls after the manipulation, according to Mann Whitney U-tests. From "Do female red-winged blackbirds limit their own breeding densities?" by W. A. Searcy. *Ecology* 69:85–95. Copyright 1988 by the Ecological Society of America.

crystalline testosterone packed into lengths of silastic tubing and implanted under the skin of the back. Control females were given empty implants. Treated females gave significantly higher rates of two aggressive displays, the "teer" vocalization (see chapter 9) and song spread. Settlement data were gathered for eight territories with experimental females and six with control females. If aggression of primary females limits settlement on their territories, we would expect lower harem sizes on the experimental territories. Instead, mean harem size was slightly higher on experimental territories (mean = 2.5) than on controls (mean = 2.3), although the difference was not significant.

Finally, Yasukawa (1990) performed an experiment to test whether the aggressive "teer" vocalization of female redwings affects female settlement. Working in a prairie in Wisconsin, Yasukawa studied four or five pairs of territories, matched for habitat characteristics, in each of three years. Within each pair, one territory was designated as a control and one as an experimental, and on the latter were placed a cassette tape player and two speakers, 28 meters apart. Teer vocalizations were played from the speakers for approximately five hours each morning, starting before the settlement of the first female on the study area and continuing until a female had been resident on each experimental territory for five days. Playback switched from one speaker to the other every 30 minutes, simulating a female moving between two display perches. The experiment was completed despite equipment failures and the theft of all the cassette players prior to the last year of playbacks. In 11 of 14 replicate pairs, the experimental territory was settled before the control (Table 5.1). Dates of first settlement were significantly earlier on the experimental territories than on control ones (Wilcoxon T = 19, P < 0.05). During the observation period, a mean of 1.2 females settled per experimental territory, compared to 1.1 per control territory (χ^2 = 0.636, P > 0.10). This experiment, then, does not support aggressive limitation, but rather accords with a positive effect of residents on subsequent settlement, at least in terms of the timing of settlement.

What can be concluded from this mix of experimental and observational results? With respect to the evolution of social polygyny, the most important question concerns whether resident females do or do not have a negative effect on further settlement. As always, we believe more weight should be given to the experimental evidence than to observational results. Five experiments have been done that test for both passive and aggressive density limitation. Four of the five indicate no negative effect of resident females on further settlement; these are the three removal experiments of Searcy (1988a, unpublished) and the playback experiment of Yasukawa (1990). The one experiment that supports density limitation by resident females is the removal experiment of Hurly and Robertson (1985). We conclude that the bulk of the evidence indicates that resident females do not have a negative effect on further settlement.

Table 5.1

Dates of Settlement by Females on Areas from Which
Female "Teer" Vocalizations Were Played Compared to
Dates of Settlement on Matched, Silent Control Territories.

	Date of Settlement	
Year	Control Territories	Playback Territories
1987	9 April	7 April
	9 April	12 April
	11 April	8 April
	16 April	13 April
	10 April	5 April
1988	9 April	5 April
	30 April	9 April
	8 April	31 March
	5 April	20 April
	10 April	5 April
1989	24 April	19 April
	26 April	22 April
	22 April	23 April
	27 April	21 April

SOURCE: Yasukawa (1990).
NOTE: Within each matched pair of territories, settlement tended to
occur earlier on the one with playback.

It may be the case that the presence of residents discourages further settlement in some areas (e.g., Ontario) and not in others (e.g, Pennsylvania,
Washington, and Wisconsin); however, more work would be needed to
demonstrate that consistent between-site differences exist. For now, we will
adopt the simpler conclusion that on the whole, there is no negative effect of
residents on settlement. The question of why resident females are aggressive
toward nonresidents we will discuss later (chapter 9).

Female Choice for Traits of Territories and Males

In this section we consider whether female red-winged blackbirds, in choosing where to settle, are influenced by particular attributes of territories or
territorial males. The results reviewed above on the consistency of female
preferences imply that female choice for certain attributes does occur, with
at least some agreement in preferences among females, but these results do
not tell us on what specific attributes choice is based.

In general we would expect female choice to be influenced by parameters
that influence female fitness (Searcy 1979a), because natural selection

should promote choice based on such attributes. As we have seen (chapter 4), a number of aspects of the territory can influence seasonal reproductive success, a component of female fitness, and therefore we might expect these territory attributes to influence female choice. In addition, we have seen that variation in male parental care may influence female breeding success, so the quality of male parental care might also influence choice. A final possibility is for male genetic traits to influence choice. As noted earlier, the evolution of female choice based on male genetic traits is more problematic than the evolution of choice on traits that affect female fitness more directly (Searcy 1982, Kirkpatrick and Ryan 1991). Moreover, we must keep in mind the fact that female redwings have the potential, at least, of uncoupling settlement choices from choice of male genes, by copulating with males other than the territory owner. Nevertheless, it is possible that female choice of where to settle is influenced by male genetic traits as well as by territory quality and male parental quality; in this section we examine all three possibilities.

ATTRIBUTES OF THE TERRITORY

The best way to determine whether female choice is influenced by specific attributes of the territory is to manipulate those attributes and measure any changes in female preferences. Unfortunately, such manipulations are difficult and have been attempted only rarely. Instead, most evidence on the effects of territory traits on choice is based on a less satisfactory method, in which measures of female preference are correlated with natural variation in territory traits. This method is subject to the usual pitfalls of inferring causation from correlation; in particular, a specific territory trait that is correlated with female preferences may not actually influence female choice itself, but only be correlated with some other territory trait, or even male trait, that influences choice. The danger of this pitfall can be minimized, however, if evidence can also be provided that the territory trait of interest influences female fitness; if the trait is correlated with female preference, and preference for the trait would be adaptive, this provides fairly convincing evidence that choice is based on that trait.

Perhaps the most fundamental level at which female red-winged blackbirds might exercise choice of a territory is in choosing between marsh and upland habitat. It is possible that females never make this choice but rather simply return to the habitat type in which they were raised. Few data exist to test whether females show this type of fidelity the first time they breed. If females were highly faithful to natal habitat type, and males were also, then upland- and marsh-nesting redwings might form genetically separate populations, even within the same area. This possibility could be tested by examining the degree of genetic divergence between marsh and upland populations

in the same locale. Redwing populations in general show little genetic divergence, even when widely separated geographically (Ball et al. 1988, Gavin
et al. 1991), so divergence between nearby marsh and upland populations is
perhaps unlikely. In the absence of evidence to the contrary, we will assume
that there is exchange of young redwings between upland and marsh and that
at least some females exercise choice of upland versus marsh habitat.

Overall there is little evidence of a consistent difference in female reproductive success between marshes and uplands (Table 4.4). Within specific
areas such differences can exist, however, and in these cases there is evidence that females prefer the superior habitat. For example, Robertson
(1972) examined two marshes and three upland areas in Connecticut and
found nesting success to be appreciably higher in the marshes (Table 4.4).
At the same time, nest density proved to be more than 10 times as high in
marsh as in upland habitat. In contrast, Ritschel (1985) studied one wetland
site, dominated by cattails and bulrushes, and a nearby upland site, dominated by mustards and wild radish, in California. Combining data from three
years of study, fledging success per nest was almost four times higher in the
upland area than in the marsh (Table 4.4). Rates of weight gain in chicks
were as high or higher in the uplands. The higher breeding success in the
upland area was due mainly to lower losses to predation there than in the
marsh, which reverses the usual trend. At the same time, harem sizes were
consistently larger in the upland site, with an overall mean of 6.2 compared
to a mean of 4.0 in the marsh. This difference was significant ($P < 0.01$)
using number of territories as the sample size. Case and Hewitt (1963) studied upland and marsh sites near Ithaca, New York. Here nesting success
was, on average, practically identical in the two habitats (Table 4.4), and
harem sizes were only slightly larger on average in the marshes (mean =
2.2) than in the uplands (mean = 1.8). Thus, there is some evidence that
female red-winged blackbirds show preferences for nesting in specific habitats, but only when such preferences are advantageous to them.

The next question is whether female red-winged blackbirds prefer some
territories over others within marsh or upland sites. Holm (1973) investigated this question on a series of marshes in eastern Washington State over
two years. She found that in both years harem sizes were significantly larger
on territories that contained mostly cattails than on those containing mostly
bulrushes. Fledging success was substantially higher on cattail territories in
one year but not the other. In this case there is evidence that females prefer
to settle on cattail territories and less convincing evidence that the preference
is adaptive.

Weatherhead and Robertson (1977a) argued that female redwings do not
show preferences for those territories with characteristics contributing to
breeding success. This conclusion was based on the results of a two-year
study of 10 marshes in Ontario. Weatherhead and Robertson (1977a) first

showed that the probability of nest success was related to four aspects of the nest site: water depth, nest height, type of support vegetation, and density of surrounding cover. They then formulated an arbitrary method of scoring nest sites based on these four attributes, and they averaged the resulting scores across the nests found within a territory to give a territory (or nest site) quality score. Territory quality scores were found to correlate significantly with reproductive success per female but not with harem size. One problem with this result is that the small sample size for the territory-score-versus-harem-size analysis, which was based on means rather than on individual values, provided little power for significance testing.

The most ambitious correlational study of female choice on territory quality is that of Lenington (1980). Lenington's general strategy was to formulate multiple regression equations that predicted components of female reproductive success based on characteristics of the territory and then test whether predicted success correlated with measures of female preference. The components of breeding success on which she focused were the probability of nest success (i.e., the probability that the nest would produce any young) and the number of young fledged per successful nest, as well as their product, which gives the number of young fledged per nesting attempt.

For one marsh in New Jersey, Lenington found that the probability of nest success was an increasing function of the density of cattails, whereas the number of young per successful nest was an increasing function of territory size and the amount of cattail edge, and a decreasing function of harem size. Multiplying her two regression equations together gave her an equation predicting number of young fledged per nesting attempt. Predicted success based on this latter equation gave a small, nonsignificant correlation with order of settlement by 18 females on 13 territories. The predicted number of young per successful nest, however, was significantly correlated with order of choice (Kendall's tau = 0.57, P < 0.01). Lenington argued that because cattail density was negatively correlated with the amount of cattail edge, it would be difficult for females to maximize both the probability of nest success and the number of young fledged per successful nest simultaneously. Still, it was difficult to explain why females seemed to choose on traits predicting number of fledged per successful nest, when the probability of nesting success was much more important to overall breeding success.

The analysis using settlement order to measure female preferences utilizes only a small part of Lenington's data, however. Much larger sample sizes, including new study sites, are available if instead harem size is used to measure female preferences. For four marshes in Illinois, Lenington found that cattail density was the only significant predictor of overall nesting success, and on these sites harem size was strongly correlated with cattail density (Figure 5.8). Harem size was also significantly correlated with territory size, although territory size was not a predictor of nesting success. In New

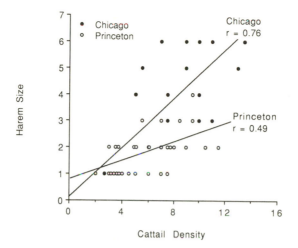

Figure 5.8 Relationship between cattail density (stems/square meter) and harem size for marshes in Illinois and New Jersey, from Lenington (1980). Data from Illinois are shown as solid circles, data from New Jersey as open circles. In both cases there was a significant, positive relationship between harem size and cattail density.

Jersey, where Lenington had two years of data on one marsh, cattail density and amount of cattail edge were significant predictors of overall nesting success. Harem sizes were positively correlated with cattail density (Figure 5.8) but not with amount of edge. We conclude that females in both New Jersey and Illinois show adaptive preferences among territories based on cattail density, though in neither area are their preferences optimal.

Whittingham and Robertson (1994) provided some correlational evidence that food availability affects female settlement patterns, but better evidence comes from two experimental tests. Ewald and Rohwer (1982) manipulated food abundance by providing extra food on marsh territories in Washington State. The food provided was a mixture of sunflower seeds, cracked corn, and commercial dog food. In the first year of the study, Ewald and Rohwer placed a single feeder on each territory and began providing food on either 31 March (eight territories) or 5 April (five territories). One problem with the experimental design was that experimental and control territories were segregated, with experimental territories on two marshes and control territories on four others (see Hurlbert [1984] for criticisms of such designs). Results were that harem sizes averaged 4.7 on experimental territories versus 3.4 on control territories (P < 0.05). In an attempt to control for the possibility that the difference in harem sizes was due to intrinsic differences in the control and experimental marshes, Ewald and Rohwer compared the ratios of numbers of females nesting during feeding to numbers nesting before feeding. These ratios were significantly higher for experimental territories

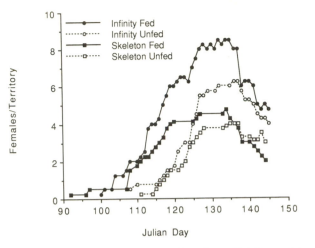

Figure 5.9 Average number of females per territory for artificially provisioned ("fed") and control ("unfed") areas on two marshes (Infinity and Skeleton) in Washington, from Wimberger (1988). Harem sizes tended to be greater on fed territories in both years, especially early in the season. Maximum harem sizes were significantly larger on fed than unfed territories on Infinity Marsh, but not on Skeleton.

(mean = 3.8) than for control territories (mean = 2.0, P < 0.02 one-tailed). Food manipulation was repeated in a second year, but because the males owning experimental territories were also removed that year, the results on female settlement are difficult to interpret.

Wimberger (1988) performed another feeding experiment, again in Washington State. Wimberger provided one feeder per territory and stocked the feeders with sunflower seeds starting on 4 April. Two marshes were used, each containing both experimental and control territories, but unfortunately experimental and control territories were again segregated, though this time within rather than between marshes. Harem sizes tended to be larger on experimental territories on both marshes (Figure 5.9); the results are convincing, despite the weakness in experimental design.

A second kind of experimental habitat alteration was performed by Yasukawa et al. (1992b). These authors chose 14 blocks, each 40 × 80 meters, in parts of their upland habitat lacking high vegetation. Blocks were divided into halves, and one half was randomly chosen for the experimental treatment. The other half served as a matched control. A single, two-meter-high pole was placed in the center of each experimental block, thus providing an elevated perch. Subsequently, female red-winged blackbirds initiated nesting attempts on 13 of 14 experimental areas, compared to only seven of 14 controls (P < 0.05 by a chi-square test; see Figure 5.10). Independent evidence suggests that the female preference for areas with perches

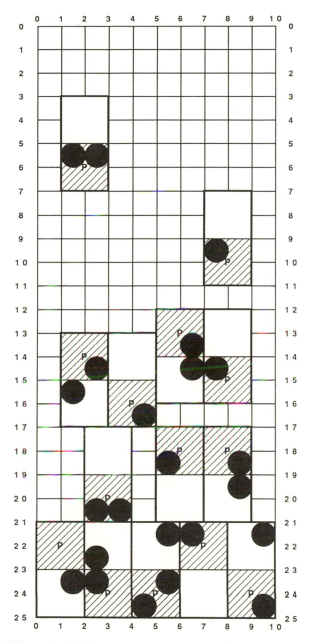

Figure 5.10 Effects of artificial perches on settlement by female red-winged blackbirds on an upland area in Wisconsin, from Yasukawa et al. (1992b). Hatched squares are experimental blocks, each of which contained an artificial perch (P) at the center. Open squares are matched control blocks. Solid circles indicate locations of nests built after perches were erected. Experimental blocks were significantly more likely to attract one or more nests than were control blocks (P < 0.01 according to a chi-squared test of independence).

is adaptive, in that nests near artificial or natural perches experience higher success (chapter 4).

Another habitat alteration, of a type, was performed by Erckmann et al. (1990), who added old redwing nests to seven 40-meter-long strips of marsh in each of two years. This experiment was conceived as a test of female copying, though it is also possible that old nests might attract settlement because they directly affect nesting success, perhaps by confusing predators. At any rate, in the first year of the experiment, a mean of 9.7 nests per plot were built on the seven plots with added old nests, compared to a mean of only 6.1 on control plots; this difference was significant ($P < 0.05$). In the following year, however, the control and experimental plots were reversed, and plots with added old nests actually had fewer new nests built on them (6.7 nests/plot), on average, than did controls (mean = 8.4), though this time the difference was not significant ($P > 0.05$). Overall, then, there is no clear effect of old nests on female settlement.

We conclude that female red-winged blackbirds probably show preferences based on habitat type, preferring whichever type has higher nesting success within a given area. Within a single habitat type, females show finer preferences based on territory attributes that influence their breeding success. Correlational evidence suggests that such attributes may include vegetation type, as in Holm's (1973) study, and vegetation density, as in Lenington's (1980). The experiments manipulating food density on territories are flawed, but they do provide some support for an influence of food abundance on female choice. A better-designed experiment has shown that females prefer nesting in areas with elevated perches.

MALE PARENTAL BEHAVIOR

Studying the effects of male parental behavior on female preferences is even more difficult than studying the effects of territory quality. For one thing, manipulating male parental behavior is especially difficult. An appropriate manipulation would have to produce a change in male parental behavior that is assessable by females when they settle, and do so without altering other characteristics of the male and his territory. No one has attempted such an experiment. As for the correlational approach, we are again faced with the problem that any correlation that we find between female preferences and a male trait might occur not because the particular trait influences female choice, but because it is itself correlated with some other male or territory trait that influences choice. There is the additional problem that a correlation between male behavior and female preference might occur because female preferences change male behavior, rather than vice versa.

Quite apart from the problems we may experience in demonstrating female choice on paternal behavior, female red-winged blackbirds face a difficult task in exercising such choice. In general we expect females to base their preferences on traits that influence their fitness, that vary between males, and that are assessable at the time of choice (Searcy 1979a, 1982). Paternal care meets the first two criteria (chapters 2 and 4), but does it meet the third? The problem is that paternal care is expressed directly only after the female has settled on a territory and begun nesting, so if paternal care is to influence a female's settlement decisions, she must have some indirect means of assessing paternal care prior to settlement. Therefore, the issue of the assessability of paternal care is an important one.

The two major aspects of paternal care are provisioning of young and nest defense. We have already presented evidence that both provisioning effort and nest defense vary between males (chapter 2) and that both affect female reproductive success (chapter 4). As for assessability, we have also presented evidence that provisioning effort increases with breeding experience among males (chapter 2), so females ought to have some basis for assessing future male provisioning effort if they can assess male experience at the time of settlement. Assessment of male experience ought to be possible. One cue to breeding experience in males is song repertoire size, which increases with age in males (see chapter 8). A second cue to breeding experience is courtship intensity. We measured the amount of time spent by males in courting and chasing females (per female in the harem) for territory owners in Yasukawa's Indiana population. We found that courtship time was significantly positively correlated with number of years of breeding experience ($r_s = 0.489$, N = 54 male-years, P < 0.01; Searcy and Yasukawa 1981b). Moreover, courtship time was significantly, positively correlated with time spent by males in feeding nestlings (per time on territory per nest with nestlings four days or older; $r_s = 0.690$, N = 39, P < 0.01). Thus, this study indicates that male breeding experience, and especially courtship time, is a good predictor of future male provisioning effort. Similar results were obtained by Eckert and Weatherhead (1987b) in a study of an Ontario population.

Two studies have addressed the question of the assessability of future male nest defense. Yasukawa et al. (1987b) measured the intensity of courtship shown by male redwings in Wisconsin toward a taxidermic mount of a female, presented in the period just before females arrived on the breeding sites. Later they rated nest defense shown by males toward a model crow presented at each male's primary nest. For 38 males, the correlation between courtship intensity and intensity of nest defense was nonsignificant ($r_s = 0.221$, P > 0.05). In the second study, Eckert and Weatherhead (1987b) rated the aggressive response of male redwings shown toward a human observer approaching a nest. Some males were also tested for aggressiveness shown toward a crow model. Male courtship intensity again proved to be a poor predictor of nest defense. The only male trait that was consistently correlated with nest defense, in various groups of males, was epaulet length, measured as the distance from the bend of the wing to the farthest extent of yellow plumage.

We conclude that both aspects of paternal care may be predictable before settlement: provisioning effort from male experience and/or courtship intensity, and nest-defense effort from epaulet length. This suggests that choice on paternal quality is possible; we next consider evidence on whether such choice occurs.

If female choice was based, in part, on male provisioning, we would expect correlations between female preferences on the one hand and male feeding effort, breeding experience, and courtship intensity on the other. Weatherhead and Robertson (1977b) rated the intensity of courtship shown toward a freeze-dried mount of a female red-winged blackbird by 11 males on an Ontario marsh. These authors found a significant, positive correlation between display score and the number of females settling per unit area of territory ($r_s = 0.87$, $P < 0.01$), which was taken as evidence of female preference for intensely courting males. Yasukawa (1981a) found that harem size in an Indiana population was significantly positively correlated with male age ($r_s = 0.486$, $N = 25$, $P < 0.01$), with number of years of breeding experience ($r_s = 0.706$, $N = 54$, $P < 0.01$), and with courtship time per female ($r_s = 0.489$, $N = 54$, $P < 0.01$). Results of both these studies are in accord with female choice based on provisioning effort. Weatherhead (1984), however, has criticized the conclusion that females choose in part on future provisioning effort. Weatherhead points out that in the Indiana population studied by Yasukawa (1981a), provisioning males assisted at only 32 of 94 nests (34.4%), with their assistance biased very strongly toward primary females (Patterson 1991). Therefore, for females settling after the primary female, there would usually be no benefit to preferring provisioning males. Moreover, in this Indiana site, harem sizes were not significantly larger for provisioning males than for nonprovisioning (Patterson 1991). Finally, harem sizes increased with age in at least one western population (J. Picman cited in Weatherhead 1984), even though provisioning by males seems to be very rare in the West.

Despite Weatherhead's criticisms, we believe choice on expected provisioning effort is still a viable hypothesis. The fact that provisioning males do not always obtain more females than nonprovisioning males is not unexpected, given that females cannot directly identify provisioning males at the time of choice; what is more important is that female preferences do correlate with those factors that predict future provisioning. The argument that most females choosing provisioning males should know in advance that they will receive no help is more of a problem for the hypothesis; however, other studies of male provisioning in red-winged blackbirds have found that male help is more evenly distributed among harem females than in Patterson's

(1991) Indiana study, with many females of lower settlement status receiving assistance (see chapter 2). Even in populations in which males strongly favor their first mates, there is a reasonably good chance that broods of later-settling females will be provisioned by the male (Yasukawa et al. 1993). Finally, we believe that the increased mating success of older males can be explained by female preferences based on provisioning even in western populations, because male provisioning, even though rare, is more likely in older males and does enhance female reproductive success in such populations (Beletsky and Orians 1990).

Nevertheless, we agree with Weatherhead (1984) that choice based on male provisioning is unproven and that there are alternative explanations for the mating advantage of more experienced males. The alternative favored by Weatherhead (1984) is that females prefer experienced males because such males, by surviving and holding territory, have demonstrated their possession of good genes. Another alternative is that experienced males own better territories and have higher mating success because their territories attract more females. This last alternative is supported by evidence that male territory size increases with increasing male breeding experience (Yasukawa 1981a).

If females chose males on the basis of traits that predict nest-defense effort, we would expect correlations between females' preferences on the one hand and male nest-defense effort or epaulet size (which predicts nest defense) on the other. Searcy (1979b) examined correlations between epaulet length and harem sizes using data gathered on several marshes in Washington State over three years. Despite large sample sizes (42, 34, and 41 males), none of the within-year correlations were significant, nor were the correlations consistently positive. Yasukawa et al. (1987b) correlated nest-defense intensity to harem size and found a significant positive correlation for males holding territories in marshes, but not for males in prairie habitat.

We conclude that the problem of assessing future paternal care is not as great as one might think; instead, there do seem to be cues allowing both future provisioning effort and future nest-defense effort to be estimated prior to mating. Evidence that females really do choose where to settle based on such cues is, however, scanty.

MALE GENETIC TRAITS

By choice on male genetic traits, we mean that females express preferences for male attributes that are in part genetically inherited, with the further condition that the preferences have evolved because of their effect on the genetic makeup of offspring. The strategy adopted in searching for this type of preference depends on the attitude one takes in the debate between the good-genes and Fisherian schools. If we view the good-genes school as correct, then we can confine our search for preferences to those male traits that are correlated with viability. If, however, we view the Fisherian school as correct, then preference for almost any male trait is a possibility. Under certain uncorrelated (with viability) choice models, the preferred trait becomes greatly exaggerated in a "runaway" process (Fisher 1930), so it might be a good strategy to focus our search on seemingly exaggerated male traits; however, exaggeration is not a necessary consequence of all such models (Kirkpatrick 1987), so female preferences for uncorrelated (with viability), nonexaggerated male traits are also possible.

Searcy (1979b) examined correlations between harem size and a variety of traits of territorial males, such as wing length, weight, condition, epaulet size, epaulet color, song rate, flight-display rate, and breeding experience. Three years of data were analyzed, all from Washington State. Strong positive correlations between male traits and harem sizes could be taken as evidence for choice based on male traits, although other interpretations would be possible, for example that females choose on territory traits, and certain male traits aid in competition for preferred territories. At any rate, the overall impression from Searcy's (1979b) results is that few, if any, male traits are consistently correlated with harem sizes. Rather, observed correlations tended to be low, and usually nonsignificant, despite sample sizes in the range of 34 to 42 males. Correlations that were significant in one year often had a reversed sign in another. Thus, this study provides little support for an effect of male traits on female settlement.

Yasukawa et al. (1980) focused their attention on a single male trait that is arguably exaggerated, namely song repertoire size. Male red-winged blackbirds have repertoires of between two and eight song types (Yasukawa 1981b), which is not extreme compared to species such as the northern mockingbird with hundreds of song types (Derrickson 1987), but which can be considered exaggerated relative to the minimum of one song type per individual found in many species. For an Indiana population, Yasukawa et al. found a significant, positive correlation between harem size and repertoire size. Repertoire size, however, was also correlated with years of breeding experience in males, and when breeding experience was controlled statistically, the correlation between repertoire size and harem size was reduced to a nonsignificant level. Thus, this study does not provide clear evidence of a link between repertoire size and female preferences shown in settlement. The question of female preferences based on repertoire size will be considered further in chapter 8.

Weatherhead (1990b) investigated female preferences based on a male trait that could serve as an indicator of good genes, namely parasite resistance. This work was prompted by Hamilton and Zuk's (1982) suggestion that parasite resistance in males is a particularly likely target for female preferences. A general problem with choice on genes for high viability is that one expects selection to minimize genetic variation in viability, thus diminishing the heritability of viability differences and the benefit that females would obtain for their offspring by choosing mates of high viability. Hamilton and Zuk proposed that parasite resistance does not follow this scenario of decreasing heritability because coevolutionary changes in the parasite and host act to maintain genetic variability in parasite resistance among hosts. Consequently, preferences for parasite-resistant males are likely to continue to be advantageous to females.

Hamilton and Zuk (1982) also argued that parasite resistance should be readily assessable by females. Parasite infection may limit the development of display structures and the ability to perform display behaviors, and thus such structures and behaviors may serve as external cues of parasite resistance. Weatherhead (1990b) tested this idea for red-winged blackbirds by comparing males infected with blood parasites to uninfected males in terms of three types of traits: (1) morphological measurements, (2) courtship behaviors shown toward female models, and (3) aggressive behaviors shown toward male models. A discriminant function based on all such traits correctly classified 54 of 67 males (80.6%) into the categories infected and uninfected. The aggressive behaviors shown toward male models were the most important components of the discriminant function, however, and these behaviors may not be readily observable by females prior to mating. A discriminant function formed without the aggressive behaviors correctly classified only 42 of 67 males (62.7%), which was not significantly better than chance.

Weatherhead et al. (1993) examined correlations between male traits and parasite burdens for a much wider range of parasites, including gut and external parasites as well as blood parasites. Little association was found between parasite burdens and any male trait, including behavioral responses to models of males and females. Moreover, burdens of the different types of parasites were for the most part independent of one another, indicating that level of parasite infection is not a unitary character. These authors also investigated the association between parasite burdens and plasma testosterone level, as it has recently been proposed that testosterone mediates the relationship between parasite loads and display traits through a feedback loop, with parasite loads depressing testosterone levels, and testosterone stimulating display traits (Folstad and Karter 1992). Weatherhead et al. (1993) did not find the predicted negative correlation between testosterone and parasite loads.

Weatherhead (1990b) found no evidence that unparasitized males were preferred by females over parasitized ones. The mean harem size of 52 parasitized males over three years was 2.06 compared to a mean of 2.16 for 49 unparasitized males; the difference was not significant (Mann Whitney U test, z = 0.435, P > 0.10). Furthermore, there was no correlation between harem size and the discriminant function score predicting whether males were parasitized ($r_s = 0.008$, $P > 0.10$).

In sum, there is not much evidence that settlement choices of female redwinged blackbirds are influenced by male traits offering only an indirect benefit to females, that is, a benefit experienced only through the female's offspring.

Conclusions

Both theory and observation support the conclusion that female red-winged blackbirds exercise choice of where to settle. Theory suggests that male redwinged blackbirds should be rather nondiscriminating in accepting females onto their territories, and there is little evidence contradicting this expectation. Among the strongest evidence for discrimination in females are the results on consistency in female choice, showing that different groups of females, or the same females choosing at different times, tend to rank the same territories in the same way. A number of experiments have investigated whether the presence of resident females on a territory affects settlement by additional females; the bulk of the evidence indicates that resident females do not have a negative effect on further settlement. Finally, there is fairly convincing evidence that settlement is influenced by specific attributes of the territory, such as habitat type, vegetation type and density, food abundance, and presence of elevated perches. There is as yet little convincing evidence that male attributes of any type affect settlement preferences.

The factors that determine female choice of breeding situation, and thereby female settlement patterns, have attracted as much attention from redwing enthusiasts as any other topic. Despite the tremendous time and effort that have been expended in attempting to understand why females choose particular territories, more work is needed. Again, we make a plea in particular for more experimental work. We have already learned much from experimental manipulations of female density, but it would still be profitable to extend such experiments to new geographic regions and habitats. A beginning has been made in manipulating territory characteristics, but certainly much more could be done along these lines, for example by manipulating water depths or vegetation densities. Manipulating the appropriate traits of territory owners is obviously going to be more difficult, but we do not believe it is altogether impossible.

6   Polygyny

We will now attempt to answer one of the two major questions addressed in this book: why are red-winged blackbirds polygynous? As we stated in chapter 1, we emphasize the social, rather than the genetic, mating system. Furthermore, we continue to make a distinction between short-term models of polygyny, which seek explanations for the present occurrence of polygyny given existing conditions, versus long-term models, which seek to explain how polygyny might have originated in the past, under hypothetical past conditions.

In this chapter we will first review some earlier attempts to use data from red-winged blackbirds to test short-term models of polygyny. We will then turn to our own tests of the set of alternative models of polygyny that we laid out in chapter 1 (see Table 1.3). We have already presented some of the data needed to test these models in the preceding chapters; here we will combine these data with new evidence to make the logical sequence of tests. After we have arrived at a short-term explanation of polygyny, we will turn more briefly to speculation on long-term models. Finally, we will consider a relatively recent object of study, the genetic mating system.

Previous Tests of Polygyny Models for Red-winged Blackbirds

As was true in most earlier work on polygyny, initial tests of polygyny models for red-winged blackbirds tended to concentrate on a single hypothesis, the polygyny threshold model (PTM). As one example, Orians (1972) performed a number of between-species and within-species tests of the PTM using icterines. One within-species prediction he tested was that "breeding success per female should *not* decrease with increasing number of females per male" (emphasis in original). Orians tested this prediction using data from a study of red-winged blackbirds in Oklahoma by Goddard and Board (1967) and from Celia Haigh Holm's study of redwings in eastern Washington State (Haigh 1968). The Oklahoma data showed no consistent relationship between fledging success per female and harem size, whereas the Washington data showed that fledging success increased with harem size,

Table 6.1
Harem Size versus Number of Young Fledged per Nest, for Marshes in
Oklahoma and Washington.

Locale	Harem Size	Number of Nests	Fledglings per Nest
Oklahoma	1–3	29	0.55
	4–5	54	0.54
	6–7	55	1.32
	>7	105	0.79
Washington 1966	1	9	0.88
	2	37	0.78
	3	38	1.48
	4	39	1.03
	5	52	1.27
	6	22	1.61
Washington 1967	1	8	0.43
	2	40	0.70
	3	52	0.81
	4	36	0.89
	5	35	1.24
	6	6	1.83

SOURCES: Orians (1972). Oklahoma data from Goddard and Board (1967); Washington data from Haigh (1968).

more or less uniformly, in each of two years (Table 6.1). Orians concluded that the data verify the prediction.

There are two major problems with this test, however. The first is that the prediction is actually not a necessary consequence of the hypothesis; that is, it is possible to have the PTM be valid for a given species yet have the prediction not hold. This possibility was pointed out by Altmann et al. (1977), who provided graphical illustrations showing how monogamous females can have either lower or higher reproductive success than "bigynous" females (i.e., females mated to bigamous males) under the PTM assumptions. The second problem with the test is that the prediction is also fully compatible with other polygyny models. For example, any hypothesis that assumes no cost of polygyny would also predict that breeding success per female would not decrease with increasing harem sizes.

Garson et al. (1981) provided further tests of the PTM, using data from red-winged blackbirds as well as other species. One prediction tested with redwing data was that "male mating status should be negatively correlated with the order in which primary, secondary, etc. females settle." This prediction is, indeed, a logical consequence of the PTM, but it is also fully compatible with many other polygyny models, for example all models assuming that female choice is based on traits of males and/or their territories. We have already looked at the redwing data sets that bear on this prediction (chapter 5); those of Garson et al. (1981) and Orians (1980) uphold the prediction, whereas those of Lenington (1980) do not.

A second prediction tested by Garson et al. (1981) with redwing data was that "fitness gains achieved by females should be negatively correlated with their settling order, regardless of their chosen breeding status." What this means is that females choosing early should have higher reproductive success than those choosing later. This prediction is a consequence of combining the assumption of directed choice, which causes the best breeding situations to be chosen first, with the assumption of a cost of polygyny, which causes the best breeding situations to be devalued once they are chosen. The predicted trend was confirmed with Dolbeer's (1976) data on an Iowa population of redwings, and in two years out of three of Holm's study of Washington redwings. This prediction is a particularly muddy one, however. For one thing, the prediction might not hold even under PTM conditions if the breeding success of early settlers on good territories is lowered sufficiently by the settlement of later females. For another, the same trend is also predicted by models that assume a cost of polygyny but no compensation. Finally, this is a trend that might be produced under almost any polygyny model by conditions outside the model's assumptions, such as a deterioration of environmental conditions for breeding over the course of the breeding season, or a tendency for poorer quality (especially younger) females to settle later.

A final example of a test of the PTM using data from red-winged blackbirds is provided by Lenington (1980) in a study we reviewed in some detail in chapter 5. Briefly, Lenington formulated multiple regression equations that predicted female reproductive success based on habitat variables and then tested whether order of choice correlated with predicted success. Positive correlations are predicted not just by the PTM but by any model assuming female choice on territory quality. Support for the prediction was equivocal, in that order of choice was correlated with the predicted number of young per successful nest but not with the predicted number of young per female.

The major problem we see with all these earlier studies arises from the fact that they examine only a single hypothesis and therefore end by testing predictions that are fully compatible with other hypotheses as well. Confirmation of such predictions, then, advances us little toward understanding the system. The way to escape this problem is to use "strong inference" testing (Platt 1964), in which one first sets out alternative hypotheses and then devises experiments or observations that have multiple possible outcomes, each of which will exclude one or more of the hypotheses. In the next section we apply this strategy to explaining the occurrence of polygyny in red-winged blackbirds.

Testing Alternative Models

MALE COERCION VERSUS FEMALE CHOICE

In our hierarchical arrangement of alternative models of polygyny (Table 1.3), our first division is between male-coercion and female-choice models. A male-coercion model assumes that males force females to mate polygynously, whereas a female-choice model assumes males are unable to force polygynous mating, which therefore occurs only due to female mating decisions. This dichotomy is placed first in our scheme because if male coercion is operating, many or most of the assumptions of other models are irrelevant, especially those having to do with costs and benefits of mating decisions to females.

In red-winged blackbirds, male coercion would require that males force females to remain on their territories to breed, which seems to us impossible. Males in certain species of polygynous, territorial ungulates, such as waterbuck (Spinage 1982) and Grant's gazelles (Walther et al. 1983), do attempt to force females to remain on their territories by "herding" them away from boundaries. Conditions are more favorable to the success of such efforts in male ungulates than in male redwings in several respects. For one, female ungulates often group of their own accord, which greatly facilitates herding (Kitchen 1974). Female redwings seldom group during the breeding season. In addition, female redwings are more mobile than female ungulates, in that they can move in three dimensions rather than two, and can probably reach a territory boundary faster. Even with their advantages, male ungulates are not very successful in their herding attempts (Kitchen 1974, Jarman 1979, Spinage 1982).

If there is little reason to think coercion would be successful in red-winged blackbirds, there is also little evidence that such behavior is attempted by males. The one regularly observed behavior that might be interpreted as male coercion is sexual chasing. As discussed in chapter 5, sexual chasing involves a male flying in pursuit of a female, and usually striking or grabbing her if he can. This description of the behavior, together with the fact that the object of pursuit is usually a female that is a resident on the pursuer's territory, is consistent with an interpretation of the behavior as male coercion, in the above sense. Sexual chasing does not, however, seem to be precipitated by the female's attempting to leave the territory; rather, males typically begin chasing a female when she is perched quietly on the territory, is gathering nest material, or has just returned to the territory from elsewhere (Nero 1956a). Moreover, pursuit often drives the female off the pursuer's territory, and neighboring males often join the chase (Nero 1956a), which are certainly not the desired effects of the type of coercion we are discussing. We conclude that neither sexual chasing, nor any other frequent male behavior, represents male coercion, and hence we reject male-coercion models for red-winged blackbirds.

Cost versus No Cost: Benefit versus No Benefit

Once the male-coercion models are rejected, the next decision is between cost and no-cost models (Table 1.3). To avoid confusion about what we mean by "cost," we draw a distinction between "component costs" and "net cost." Both types of cost refer to reductions in fitness incurred by females because they choose males that are already mated. A component cost is a loss in any single fitness component, whereas net cost is a loss in total fitness summed over all components. Net cost can be calculated as the expected fitness of a female settling with an already-mated male minus her expected fitness if she could obtain the same male as an unmated male.

Cost models assume there is a net cost of polygyny, whereas no-cost models assume there is not. If there is a net cost of polygyny, then, other things being equal, female fitness ought to decline as harem size increases, other things in this case being the quality of the male, of the female, and of the territory. As this question of a net cost of polygyny is so important in discriminating between the alternative models of polygyny, we will go into it in considerable detail.

The methods used for testing for a net cost of polygyny will also yield tests for a net benefit of polygyny. A net benefit of polygyny occurs if a female choosing a particular territory and male would have a higher expected fitness if one or more females were already resident on that territory than if she were the sole female to settle there. We will consider both the cost versus no-cost, and the benefit versus no-benefit, dichotomies in this section.

The most likely component cost of polygyny in red-winged blackbirds, as in birds in general, stems from competition among polygynously mated females for their mate's contribution to parental care (Orians 1969a). In chapter 2 we discussed the fact that some forms of care are nonshareable, in that care given to one brood reduces the care given to others. We also concluded that of the two major forms of parental care provided by male red-winged blackbirds, provisioning of young is nonshareable, whereas guarding of eggs and young is shareable. Thus, competition between females is likely only for provisioning. The magnitude of this component cost may be considerably lower in red-winged blackbirds than in most altricial birds. For one thing, male provisioning is rare in many redwing populations, specifically those in western North America (chapter 2). Furthermore, even in populations in which many males provision, some males fail to provision at all, provisioning males bring food to only a subset of their broods, and they bring food to those broods considerably less often than do females (chapter 2). The upshot is that in competing for male provisioning, females are competing for a resource that is not as valuable to begin with as it is in most species. Nevertheless, it has been shown that male provisioning has a positive effect on female reproductive success in red-winged blackbirds (chapter 4), so competition for male provisioning may have a cost to females, even if a relatively small and variable one.

The rules by which males allocate their help among females constitute another factor affecting the magnitude of the component cost of polygyny produced by competition for male provisioning. If all or most of the help is given to primary females, as is the case in certain other polygynous species, this increases the cost to those females choosing to mate polygynously. In redwings, males have been found to favor broods of primary females in some studies, but in others they have allocated provisioning quite evenly among females (chapter 2). Even-allocation reduces the cost both because females that choose already-mated males have some chance of obtaining male assistance, and because females that choose to mate monogamously may receive reduced help if other females join the harem afterward.

Sufficient data are available to allow us to estimate the component cost of polygyny due to reduced male assistance in provisioning for a marsh population in Ontario (Muldal et al. 1986) and an upland population in Wisconsin (Yasukawa et al. 1990); the calculations are shown in Box 6.1. In Ontario there was actually a benefit of secondary rather than primary status, which occurred because young of secondary females were slightly more likely to be fed by males than were young of primary and monogamous females. In Wisconsin the component cost of settling with secondary rather than primary status was on the order of one-fourth of a fledgling. The component cost should be higher or lower in other eastern populations, depending on the pattern of allocation of paternal care (Box 6.1). The component cost must be very low in the far western population studied by Beletsky and Orians (1990) because the probability of receiving any male help was so low there.

Another possible component cost of polygyny stems from competition for resources provided by the territory. In redwings, territories provide two types of resources to resident females, food and nest sites. In many populations, young are fed primarily on emerging aquatic insects (Snelling 1968, Orians 1980), which may be obtainable on the territory. For example, Orians (1980) found that 70% of the food brought to young in a Washington population were prey of aquatic origin. Females in this population, however, often nested in marshes with little or no insect emergence and flew to nearby marshes or lakes to obtain food. If females are able to forage effectively off the territory, then competition for food produced on the territory cannot be crucial. In our present study populations in Pennsylvania and Wisconsin, we have found that females feed off the territory more often than not, especially when obtaining food for their young, and the same pattern has been noted in

BOX 6.1 CALCULATING THE COST OF POLYGYNY
DUE TO LOSS OF MALE PROVISIONING

The goal here is to calculate that part of the cost of polygyny due to any decrease in the male's assistance with feeding the young. This cost cannot be estimated by simply comparing the reproductive success of, say, monogamous and secondary females because such a comparison would be confounded by any other costs of polygyny, any benefits of polygyny, and any systematic differences between monogamous and secondary females in traits other than male help, such as age or time of settlement. Instead, we seek to calculate this specific cost from data on the amount of male help obtained by females of differing status, and on the effect of male help on female reproductive success.

The method is as follows. Let p_x be the probability of obtaining male assistance for a female of status category x. Let F_a = number of fledglings raised in assisted broods (i.e., broods fed by both male and female) and F_n = number raised in broods that were not assisted (i.e., broods fed by the female only). Then $F_a - F_n = b$, where b is the benefit of being assisted by the male. Let C_{yx} be the cost due to loss of male provisioning from settling as a female of status y rather than status x; then:

$$C_{yx} = p_x(b) - p_y(b) = (p_x - p_y)b.$$

Muldal et al. (1986) give all the data necessary to calculate costs using this procedure. Their data on the proportion of females receiving male help versus female status are given in Table 2.3. The mean number of fledglings produced was 2.04 for 47 assisted nests and 1.33 for 52 unassisted nests; therefore b = 2.04 - 1.33 = 0.71 fledglings. The figures for the costs then become C_{21} = -0.08, C_{32} = 0.27, and C_{43} = 0.26. Note that the "cost" of settling as a secondary rather than a primary female is negative, meaning that there is actually a benefit to being secondary; this of course occurs because secondary females were slightly more likely to be assisted than primary and monogamous females in Muldal et al.'s study. Costs are additive, so that C_{31} = C_{21} + C_{32} = -0.08 + 0.27 = 0.19.

There are also sufficient data to estimate costs for Yasukawa's Wisconsin upland site. The mean number of fledglings produced by male-assisted broods was 2.76, compared to 1.93 by unassisted broods; therefore b = 0.85 fledglings (Yasukawa et al. 1990). Using the data in Table 2.3 on the proportion of females recieving male assistance, the figures for costs are C_{21} = 0.25, C_{32} = 0.08, and C_{43} = 0.21.

Whittingham (1989) and Searcy (unpublished) provide data on the proportion of females of differing status that get male assistance, but they do not provide the proper data on b, the benefit of male assistance. It can be seen (Table 2.3), however, that in these two studies the proportion of females helped is either essentially constant across differing categories of status (Searcy unpublished) or is actually higher for females of lower status than for primary females (Whittingham 1989). Therefore there can be little or no cost of polygyny due to loss of male provisioning for these populations.

other areas (e.g., Hurly and Robertson 1984, Whittingham and Robertson 1994). We conclude, then, that there may be some cost of polygyny due to competition for food, but this cost is unlikely to be large.

Competition for nest sites might be important if there were a limited number of good sites per territory and the first-settling female usually chose the best site, leaving all subsequent females with poorer ones. Yasukawa et al. (1992b) speculated that first settlers might seek out nest sites that males could guard effectively, because of the proximity of elevated perches, and that later settlers might be relegated to more dangerous sites. Lenington (1980) found that early-settling females tended to obtain nest sites in deeper water than late-settling females on the same territories. Still, it seems to us that even though nest-site quality may vary between areas within territories, a given nesting area usually provides numerous possible sites which must be of similar quality, so that nest sites should seldom be a limiting resource within a territory.

Another possible source of a component cost of polygyny is behavioral interference between females. Females do seem to interact more with members of the same harem than with residents of other territories, and many of these interactions are aggressive. Females appear to ignore each other most of the time, however, especially once they have begun nesting. Thus, it is difficult to assess the cost of behavioral interference. We will discuss interactions between females in more detail in chapter 9.

Veiga (1990) showed that female house sparrows mated to the same male sometimes destroy each others' eggs and young; where such behavior occurs, it produces an obvious cost to polygynous mating. Searcy (unpublished) tested whether intraspecific infanticide occurs in red-winged blackbirds by placing unguarded nests containing eggs onto territories on which female redwings were resident; similar methods had revealed intraspecific infanticide in marsh wrens (Picman 1977a, b). Trials were conducted on three territories, and the nests were observed for a total of 23 hours. Resident females often approached within one meter of the experimental nests, and one female actually perched on the lip of a nest, but no attacks were made on the eggs. We know of no evidence from any other study suggesting female infanticide in redwings, so we conclude that this cost is highly unlikely.

A final possible component cost of polygyny is infertility, the idea being that sperm depletion might be more likely in males the larger their harems, so that infertility would correlate positively with harem size. In red-winged blackbirds, infertility of eggs from whatever cause is rare, with an incidence of about 1% (chapter 4). The incidence of infertility is so low that even if infertility does increase with harem size (which is unknown), infertility is unlikely to produce a significant cost of polygyny.

It is clear that there are factors tending to produce component costs of polygyny in red-winged blackbirds, notably competition for male provisioning. The question of whether there is a net cost, however, is considerably complicated by the possibility that there are benefits of polygyny as well as costs. In red-winged blackbirds, the major component benefit of polygyny seems to be increased safety from nest predation. We have seen that females as well as males defend nests against predators (chapter 2) and that nests built in close proximity to nests of other females are safer from predation (chapter 4). The increased safety seems to be due to a mixture of joint defense and predator dilution, but regardless of the mechanism, increased safety from nest predation does seem to provide a benefit of polygyny in red-winged blackbirds.

A second possible component benefit of polygyny stems from increased foraging efficiency. Polygynous mating could raise foraging efficiency if the harem functioned as an "information center" (Ward and Zahavi 1973), with females acquiring information on the location of food by observing those harem members that have been successful in foraging. There is evidence that information transfer occurs at breeding colonies of cliff swallows (Brown 1986) and ospreys (Greene 1987). Female red-winged blackbirds do not depart the territory synchronously when foraging, nor do they consistently depart and arrive from the same directions (Gordon H. Orians and Les D. Beletsky pers. comm.), which is evidence against the use of the harem as an information center. Thus, this component benefit seems unlikely.

In examining the net cost or benefit of polygyny, the number of young fledged per female is usually employed as the fitness measure. This measure of seasonal reproductive success is not, of course, a complete measure of fitness, but it does sum over most of the possible component costs and benefits outlined above.

Studies that relate seasonal reproductive success to harem size can be either observational, utilizing natural variation in harem sizes, or experimental, examining harem sizes that are manipulated by the researcher. Of the two types of study, we believe the experimental is much preferable. The major problem with observational studies is that they do not control factors other than harem size that may influence female fitness. For example, Orians (1972) showed that reproductive success per female increased with harem size in Haigh's (1968) data (Table 6.1). This trend could be taken as evidence for a net benefit of polygyny. The trend could also be explained, however, by assuming there is a net cost of polygyny and that females do not settle polygynously unless more than compensated by a superior territory and/or male; this latter interpretation is the one made by Orians (1972). Thus, the problem with this comparison as a test for a net cost of polygyny is that one has not controlled for male and territory quality.

Conversely, suppose one found that polygynously mated females have lower reproductive success, on average, than monogamous females. This trend could be interpreted as support for a net cost of polygyny. The trend might also come about, however, because secondary females are of lower quality, perhaps younger as Crawford (1977a) found, or because they settle at worse times than monogamous females. Here the problem with the original comparison is that one has not controlled for female quality or settlement date.

A properly designed experiment can circumvent most of these problems. If harems are randomly designated as controls and experimentals, then there ought to be no systematic difference in mean territory quality, male quality, or female quality between the two groups. If the experimental manipulation is to remove females, then it will often be the case that the remaining females are a biased sample with respect to settlement date. For example, early-settling females may be removed, leaving only late-settling females on the experimental territories for comparison with late-settling and early-settling females on controls. The obvious way around this difficulty is to confine comparisons to the late-settling controls.

Two sets of removal experiments have been done with female red-winged blackbirds. In one, Hurly and Robertson (1985) did not compare reproductive success of females on control versus experimental territories; this comparison was virtually precluded by their experimental design, which used repeated removals throughout the breeding season, so that few females could have nested undisturbed on the experimental territories. Therefore, the relevant data are available only from the experiments of Searcy (1988a, unpublished). These data allow us to compare reduced versus normal harems with respect to two sources of reproductive loss, proportion of young lost to starvation and proportion of nests attacked by predators, and with respect to seasonal reproductive success per female. The probable component sources of a cost of polygyny are reduced male help with provisioning and increased competition for food, which we would expect to be reflected in higher starvation rates on control than on experimental territories. The only probable source of a benefit of polygyny is increased safety from nest predation, which we would expect to be reflected in lower proportions of nests attacked by predators on control than on experimental territories. Number of young fledged per female provides a summary measure, allowing us to look for a net cost or net benefit of polygyny.

The design of Searcy's experiments has been described previously, in the context of the effect of removals on subsequent settlement (chapter 5). In the first experiment, the first-settling female was removed from each of eight experimental territories, with 10 territories serving as controls. In the second, one to three early-settling females were removed from six experimental territories (mean = 1.7 females removed per territory), with six territories as controls. In the third, one to three secondary females were removed from seven experimental territories (mean = 1.6 females removed per territory), with seven territories serving as controls. In no case did significantly more females settle, on average, on the experimental territories after removal than settled during the same period on control territories, and in each experiment the harem sizes of females actually nesting were significantly lower on the experimental territories (Table 6.2). Thus, the manipulations were successful in changing harem sizes for the remainder of the breeding season.

As noted above, costs of polygyny are most likely to show up as higher starvation on control territories relative to experimentals. Comparing all females, starvation losses averaged higher on control territories in two experiments and higher on experimental territories in one experiment; none of these differences were significant (Table 6.2). Trends were the same when comparisons were restricted to females of comparable settlement status. Combining data from all females in all experiments, the mean proportion of nestlings starving per nest was 0.22 for 45 control nests and 0.19 for 17 experimental nests; the difference was not significant (Mann Whitney $U = 361.5$, $z = 0.378$, $P > 0.10$ one-tailed). Again, results were similar for females of comparable settlement status. There is, then, little evidence of a cost of polygyny.

Benefits of polygyny are most likely to be seen as lower predation rates on control territories relative to experimentals. Comparing all females, the proportion of nests taken by predators was identical on control and experimental territories in one experiment and lower on control territories in the other two. The difference between controls and experimentals was significant in Experiment 2. Comparing females of similar settlement status, predation was higher on control territories in one experiment and higher on experimental territories in two; again, the only significant difference occurred in Experiment 2, with higher rates of predation on experimentals. Combining the data from all females in all experiments, the mean proportion of nests taken by predators was 0.35 for 23 control territories and 0.47 for 19 experimental territories, which was not significant ($U = 179$, $z = 1.02$, $P > 0.10$ one-tailed). Using data on females of comparable settlement status within each experiment, the mean proportion of nests taken by predators was 0.25 on 20 control territories and 0.48 on 19 experimental territories, which was a significant difference ($U = 127.5$, $z = 1.85$, $P < 0.05$ one-tailed). Overall there is some evidence of a benefit of polygyny, in terms of increased safety from nest predation.

Again, our best measure of the net cost or benefit of polygyny is number of fledglings produced per female. Comparing all females, this measure was higher for females on control territories in two experiments and higher for females on experimental territories in the third. None of the trends were significant. Results were similar when comparing females of the same settlement status. Combining the data from all three experiments, 59 females on control territories produced a mean of 1.9 fledglings each, versus 1.6 fledg-

Table 6.2
Reproductive Success of Females in Harem Size Manipulation Experiments.

Reproductive Success Measures	Subjects in Analysis	Experiment 1		Experiment 2		Experiment 3	
		Controls	Experimentals	Controls	Experimentals	Controls	Experimentals
Number of Females Nesting per Territory		2.8	1.5*	2.3	0.7*	2.7	1.7*
		(10)	(8)	(6)	(6)	(7)	(7)
Proportion of Young Starving per Nest	All Females	0.30	0.27	0.07	0.25	0.15	0.04
		(25)	(9)	(9)	(2)	(11)	(6)
	Same Status	0.33	0.27	0.10	0.25	0.11	0.08
		(16)	(9)	(4)	(2)	(6)	(4)
Proportion of Nests Attacked by Predators	All Females	0.27	0.27	0.40	0.88*	0.44	0.48
		(10)	(8)	(6)	(4)	(7)	(7)
	Same Status	0.31	0.27	0.27	0.88*	0.18	0.50
		(9)	(8)	(4)	(4)	(7)	(7)
Number of Fledglings per Female	All Females	1.9	1.6	2.3	0.8	1.7	2.1
		(28)	(12)	(14)	(4)	(17)	(9)
	Same Status	1.7	1.6	1.5	0.8	2.4	2.6
		(18)	(12)	(4)	(4)	(7)	(5)

SOURCE: Searcy (1988a, unpublished).
NOTE: Experiments were conducted on marshes in Pennsylvania. Harem sizes were lowered on experimental territories by capturing and holding females.
*P < 0.05.

lings each for 25 females on experimental territories. The difference was not significant (U = 664, z = 0.751, P > 0.10 two-tailed). Restricting analysis to females of comparable settlement status, 29 females on control territories produced a mean of 1.9 fledglings each, whereas 21 females on experimental territories produced a mean of 1.7 fledglings, again not a significant difference (U = 289.5, z = 0.309, P > 0.10 two-tailed).

The data on starvation rates suggest that any component cost of polygyny through competition for paternal provisioning or directly for food must be small. There are, of course, other ways that a cost could manifest itself, notably by a decreased mass of young produced on crowded territories, but Searcy (unpublished) found no support for this cost among broods in Experiment 3. The data give better support for a benefit of polygyny through increased safety of nests from predation, though the effect was seen clearly in only one experiment of three. If such a benefit does exist, it was not substantial enough to cause significant overall differences in reproductive success between females on experimental and control territories.

The results, then, are clear in rejecting a net cost of polygyny, as the overall trend in the reproductive success data is in the opposite direction from that predicted on the basis of a net cost. The results are not as clear on net benefit, as the overall trend in reproductive success is in the direction predicted by a net benefit, but the trend is not significant. We will therefore reject both net-cost and net-benefit models for now, but we consider the rejection of net-benefit models to be more provisional.

Support for these decisions on net cost and net benefit comes from the results on settlement in the same experiments. If polygyny had a net cost, we would expect territories on which harems had been lowered to be more attractive to settlement than are control harems. As detailed in chapter 5, there was no evidence of an increase in female settlement after removals in any of the experiments, and thus this prediction of the cost models is rejected. Conversely, if polygyny had a net cost, settlement should decline on removal territories, and the settlement data showed little support for this prediction either.

RANDOM VERSUS DIRECTED CHOICE

Within the no-cost, no-benefit models, the remaining dichotomy is between random and directed choice (Table 1.3). Directed choice in this context means that females choose where to settle according to features of males and/or their territories, with the further stipulation that there be some agreement among females on what males or territories are preferred. Random choice means that there are no such preferences, or that preferences exist but with virtually no agreement among females. It is necessary that there be some agreement among females in preferences if preferences are to affect the form of the mating system.

If there is no cost and no benefit of polygyny, and if female choice is random with respect to male and territory features, then settlement is analogous to a process in which markers are cast out randomly onto the squares of a checkerboard. It should be intuitively obvious that if settlement were this random, and if there were at least as many markers (females) as squares (territories), then it should be quite common for two or more markers to end up on one square, i.e., for polygynous matings to occur. Directed choice, then, is not necessary for polygyny to occur in such a system. Directed choice does, however, have an effect on a no-cost, no-benefit system, in increasing the expected degree of polygyny.

We have already reviewed the relevant evidence on female preferences in settlement in chapter 5. This evidence supports directed choice over random choice. For example, in our computer simulations of female settlement (Yasukawa and Searcy 1986), we set up alternate models that assumed either random choice or directed choice, and we found that the directed-choice models accorded much better with observed settlement patterns. In addition, we reviewed a variety of correlational evidence suggesting female preferences for particular attributes of territories, and some experimental evidence as well. Females may also choose partially on traits of territory owners, but there is less evidence of this pattern. We conclude that the random-choice assumption should definitely be rejected in favor of directed choice.

Cost Models

Because of the way in which our hierarchy of models is organized, many of the more noted explanations of polygyny are rejected as a group at the point where a net cost of polygyny is rejected; these rejected models include the skewed-sex-ratio hypothesis, the polygyny threshold model, and the deception hypothesis. These hypotheses all depend heavily on the assumption that there is a net cost of polygyny, so it is quite legitimate to reject them when that assumption is rejected. Nevertheless, it is still valuable to consider some further tests of these models.

The skewed-sex-ratio hypothesis suggests that females are forced to pay a net cost of polygyny because of a lack of unmated males. In some studies of red-winged blackbirds, unmated, territory-owning males have been shown to be available throughout the breeding season (e.g., Orians 1973, Searcy and Yasukawa 1983), whereas in others all observed territory owners obtained at least one female (Case and Hewitt 1963, Searcy 1988a). Even if no unmated territory owners are available by the end of the season, however, it is our experience that many of the polygynous females will have chosen mated males at a time when unmated territory owners were still available. Moreover, it is also true that in populations where no territorial males remain unmated, there are still large numbers of unmated males without territories;

the evidence for this was reviewed in chapter 2. In fact, if one counts all males capable of reproduction, including nonterritorial as well as territorial males, and one-year-olds as well as adults, then there is little evidence that sex ratios are skewed at all in redwings (chapter 2). We have argued previously that nonterritorial and unmated males should be counted in assessing sex ratios in this context (Searcy and Yasukawa 1989). The skewed-sex-ratio hypothesis can thus be rejected on grounds other than absence of a cost of polygyny.

The polygyny threshold model, which we interpret as being synonymous with the class of cost-compensation models, assumes that females pay a cost of polygyny because they are compensated by acquiring a superior territory and/or mate. We have already presented evidence that in red-winged blackbirds, harem sizes are larger on territories with traits favorable to breeding success than on territories with less favorable characteristics (chapter 5). This argues that if polygyny did have a net cost in this species, polygynously mating females would indeed be at least partially compensated for that cost. Whether compensation is great enough to compensate fully for a cost of polygyny is always a difficult question to answer (Searcy and Yasukawa 1989); at any rate, the question is moot for red-winged blackbirds, as we have found no net cost. The point we wish to make here is that the only ground for rejecting the polygyny threshold model for red-winged blackbirds is the failure of the net-cost-of-polygyny assumption. If we were to be proved wrong in rejecting this assumption, in general or for a specific population, then the polygyny threshold model would be the most likely of the alternative explanations of polygyny.

Finally, we consider briefly the no-compensation models. At present, these hypotheses are rejected both because they assume a cost of polygyny and because they assume polygynous females do not obtain superior territories. In addition, the deception hypothesis is unlikely because it assumes that mated males are able to conceal their mating status from females. This hypothesis was first proposed for polyterritorial species (Alatalo et al. 1981), in which it is plausible that a female settling on a second territory might be unaware that the male had another female on another territory. The hypothesis has since been applied to a few polygynous species with single territories. For example, Catchpole et al. (1985) suggested that in the case of the great reed warbler, territories are large enough and vegetation dense enough that males can attract a second female at one end of a territory without her being aware that there is a first female at the other end. Such a hypothesis is certainly not plausible for red-winged blackbirds. Territories of red-winged blackbirds are relatively small, especially in marshes (chapter 2). Vegetation may be dense on redwing territories, but females typically spend much of their time perched high in the reeds, where they are clearly visible. In addition, females engage in vocal and visual displays that increase their conspicuousness (Nero 1956a, Orians and Christman 1968), and resident females regularly show overt aggression toward visiting females (Nero 1956b, Lenington 1980, Searcy 1986); these behaviors make the presence of resident females obvious to us, and probably to prospecting females as well. Thus, it seems impossible for newly-settling females to be unaware of the presence of a male's earlier mates.

A second no-compensation model is the search-cost hypothesis, which suggests that females knowingly accept already-mated males because the cost of searching for an unmated male is greater than the cost of polygyny. This hypothesis has been suggested for pied flycatchers (Stenmark et al. 1988), which have a rather restricted breeding season, and in which female reproductive success declines rather quickly as the breeding season progresses, providing a clear cost of extended search. Red-winged blackbirds have a longer breeding season, and reproductive success does not consistently decline through the breeding season (Caccamise 1976), so there would not seem to be the same cost to extended search as in pied flycatchers. There is little systematic evidence on the behavior of female redwings searching for settlement sites, but our impression is that females will often investigate sites for at least several days before settling. It is also our impression that it should be easy to locate unmated, territorial males within such a time frame. Therefore, the search-cost hypothesis does not seem plausible for redwinged blackbirds.

The final hypothesis in our hierarchy of models is the maladaptive-female hypothesis, which proposes that females choose males they know are mated even though more advantageous matings are available to them. This idea has little appeal to most behavioral ecologists, who tend to be either open or closet adaptationists ("running dog adaptationists," as one ethologist of our acquaintance puts it). Nevertheless, we feel this hypothesis should be kept in mind, both because it is a real possibility and because we should realize that this is the shoal upon which we will be driven if we discard all other hypotheses.

Summary of Short-term Models

In working through our set of alternative hypotheses for the explanation of polygyny (Table 1.3), our first step was to reject male coercion in favor of female choice, on the grounds that there was no evidence that male redwings attempt coercion, and that coercion would seem unlikely to be successful even if attempted. We next rejected cost models, because all component costs (such as that stemming from competition for male parental care) seem small for redwings, because there are compensating benefits (in decreased nest predation), and because the reproductive success of females in experimentally manipulated harems indicates no net cost. The experimental evidence indicates that polygyny also does not have a net benefit, allowing rejection of the benefit models. Finally, evidence reviewed in chapter 5 on female settlement patterns allows rejection of random choice in favor of directed choice. Our explanation for polygyny in red-winged blackbirds, then, is a female-choice model, in which there is neither a net cost nor a net benefit of polygyny, and in which there is directed choice based on territory quality, and possibly on male characteristics as well.

Long-term Models of Polygyny

Long-term models seek to explain how and why polygyny originated. These models are necessarily speculative because of our ignorance of the ancestral mating system from which the present one evolved and because of our equal ignorance of the environmental conditions under which evolution took place. With that caveat in mind, we present one scenario for the origin of polygyny in red-winged blackbirds in Figure 6.1. Here the redwing mating system is assumed to have evolved from what is perhaps the most typical of all passerine mating systems: the mating relationship is monogamous, males are

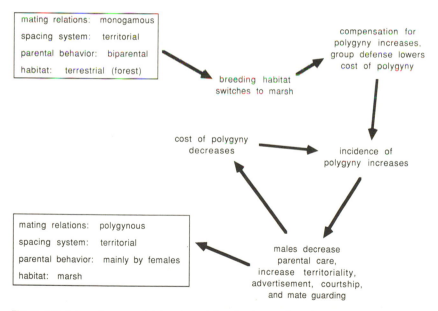

Figure 6.1 A long-term model for the origin of polygyny in red-winged blackbirds, assuming an origin from a social system resembling that of most monogamous passerines. The ancestral social system is described in the upper left box, and a hypothetical series of events then leads to the redwing social system, as described in the lower left box.

territorial in the breeding season, and parental behavior, especially provisioning of young, is shared equally between the sexes. The hypothetical ancestor is assumed to have bred in a terrestrial habitat, such as forest. The key change that triggers all subsequent changes is hypothesized to be a shift in the preferred breeding habitat to marshes. Several factors then act to make the frequency of polygyny increase. One is that the area of marshes is limited, so that a subset of the males can monopolize all the available nesting habitat within their territories (Orians 1961). If female reproductive success is higher on marshes than elsewhere, then females mating polygynously in marshes are to some extent compensated for the cost of polygyny. In addition, territories may vary in quality more in marshes than in other habitats (Orians 1969a), and this variation also can provide fitness compensation for polygynous mating. Finally, territories are smaller in marshes, and the habitat is more open, so group defense against nest predators becomes more important; this provides a benefit to polygyny that lowers the net cost. As these factors increase the frequency of polygyny, males are selected to put more effort into territory defense, advertisement and courtship, and mate guarding because the payoffs of these activities have increased. Increased effort in defense, courtship, advertisement, and mate guarding comes at the cost of decreased devotion to parental care. As male parental care decreases, the net cost of polygyny decreases, feeding back to increase further the incidence of polygyny. Out of this mutually reinforcing cycle comes our end system: mating relationships are polygynous, the spacing system is territorial, females provide the majority of parental care, and the preferred breeding habitat is mainly marsh.

One addition that could be made to this hypothesis is to assume that food is especially plentiful in marshes, which often may be the case (Orians 1980). If food availability increases as the species begins breeding in marshes, this would make it easier for females to raise young without male help, and so lower the cost of polygyny. A species could, however, respond equally well to an increase in food supply by increasing clutch size, a response that would be more advantageous to females. If clutch size increases sufficiently, then male help may remain essential to females. There would therefore seem to be a conflict of interest between the sexes concerning how to respond to increased food, with males benefiting more from increased polygyny, and females benefiting more from increased clutch sizes. It is unclear which sex ought to win this conflict, so we omit this consideration from our hypothesis.

The first place to question this model is in its choice of an ancestral mating system. Although we probably will never know what the true ancestral state was, we can get an idea about likely possibilities by examining the mating systems of closely related species, starting with other members of the genus *Agelaius*. Altogether there are 10 species in this genus. Of these, only the red-winged blackbird and tricolored blackbird are North American, the rest being from South America and the Caribbean. Many of the southern species are poorly known. Among the species whose behavior has been studied, none have a mating system similar to that of the hypothetical ancestral species in Figure 6.1.

Orians (1985) attributes specific mating systems to six of the *Agelaius* species; four of these are said to be polygynous and two monogamous. At least one of the other polygynous species, the yellow-hooded blackbird, has a mating system closely resembling that of red-winged blackbirds. Male yellow-hooded blackbirds defend territories on marshes, on which multiple females nest. Wiley and Wiley (1980) found harem sizes to be small in Trinidad and Surinam, where breeding seasons were extended, but larger in central Venezuela, where breeding seasons were short. As in redwings, males feed young less than do females, but whereas male redwings never aid in nest construction, male yellow-hooded blackbirds construct all nests by themselves, subsequently displaying at their nests to attract females. Yellow-hooded blackbirds appear to be more colonial than redwings, with breeding territories clumped within marshes, and much seemingly similar habitat unoccupied (Wiley and Wiley 1980). Grouping may provide protection against parasitism by the shiny cowbird.

The tricolored blackbird is also classified as polygynous, but it has a rather different mating system. Nesting is confined to marshes, within which breeding territories are clumped, and nesting colonies can be huge. Males defend territories for only about one week, during the time when females are building nests and soliciting copulations (Orians 1961). Territory sizes are extremely small, averaging only a little more than three square meters (Orians 1961). Harem sizes are usually two or three females per territory, seldom more (Lack and Emlen 1939, Orians 1961). Males are very active in feeding the young (Orians 1961).

One of the two *Agelaius* species known to be monogamous is the yellow-winged blackbird, which was studied by Orians (1980) in Argentina. Pairs nested in loose colonies in marshes or ditches. Orians found no sign of territoriality; both neighbors and other conspecifics were allowed near nests without aggression. Males were active in feeding young (Orians 1985).

The second monogamous species is the yellow-shouldered blackbird of Puerto Rico. Post (1981) found this species to nest in loose colonies, placed either in mangroves or in scattered trees in pastures or suburbs. Territoriality existed but was rather feeble; males defended only the immediate neighborhood of their nests, out to a distance of about three meters. All 25 pairs studied by Post were monogamous, and males participated equally with females in feeding the young. Post thought that nesting aggregations resulted from mutual attraction among the birds rather than from limited nesting habitat. Communal mobbing of nest predators was common, providing a possible benefit for aggregation.

With these *Agelaius* mating systems in mind, a new model can be pro-

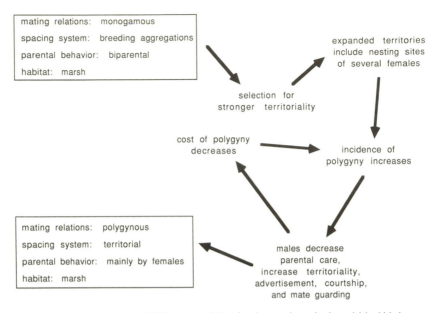

Figure 6.2 A long-term model for the origin of polygyny in red-winged blackbirds, assuming an origin from a social system resembling that of monogamous *Agelaius* species. The assumed ancestral social system is described in the upper left box, and a hypothetical series of events then leads to the redwing social system, as described in the lower left box.

posed for the evolution of the redwing system (Figure 6.2). In this model, the ancestral state is assumed to resemble the mating system of the monogamous *Agelaius* species. Here mating relationships are monogamous, breeding is done in loose colonies, territoriality is weak or absent, and parental care is shared equally by the sexes. The breeding habitat is predominantly marsh or other wetlands. The major change needed to push this system toward polygyny is an increase in male territoriality. As males devote more effort to territorial defense, territory boundaries expand, single territories take in greater numbers of nest sites, and the incidence of polygyny increases. This puts the system into the same mutually reinforcing loop as in the previous hypothesis, with the increased chance of polygyny favoring increases in male territorial and advertisement behavior at the cost of decreased parental care, the cost of polygyny consequently being reduced, and the decreased cost feeding back to increase the level of polygyny. Out of this loop comes our familiar redwing system.

In a way, the ancestral, monogamous system in this second hypothesis is biased toward polygyny to begin with, in that females are already nesting in aggregations. We assume that the original function of aggregated breeding would have been defense against nest predators, as seems to be the case in yellow-shouldered blackbirds (Post 1981). Thus, predator defense is, indirectly, again a causal factor in the evolution of polygyny. Also, the mating systems of tricolored and yellow-hooded blackbirds serve in this hypothesis as models for possible intermediates in the evolution of the redwing system, since these two species show intermediate levels of territoriality, and less extreme polygyny than redwings.

The biggest gap in this second hypothesis is that we have not specified what factor provides the original impetus toward increased territoriality in males. Territoriality in general is thought to be favored when resources are economically defensible (Brown 1964), meaning that the benefits of maintaining exclusive access to resources outweigh the costs of defense. Resources are not economically defensible both when they are too scarce and when they are superabundant. The primary resource defended by male redwinged blackbirds seems to be nest sites, with food of secondary importance. It is possible that increased territoriality became favored in redwings as the ratio of their population sizes to available nesting habitat became larger, thus making it possible for males to defend all available nest sites, i.e., all available marsh habitat. Again, this is purely speculation.

We have now presented two models of the origin of polygyny in redwings, based on two different assumptions about the form of the ancestral mating system. Many more hypotheses could be proposed, by varying the assumed ancestral state and by varying the environmental conditions assumed during the course of change. We do not have any way of testing between these first two hypotheses, so proposing further hypotheses would be superfluous. We go into the subject this far so as to make two points: first, that the question of the origin of polygyny in red-winged blackbirds is different from the question of why the system continues under current conditions, and second, that any progress toward settling the origin question depends crucially on a decision on the form of the ancestral system.

A comparative analysis of icterine mating systems, using modern phylogenetic methods, might help to resolve the issue of the ancestral mating system of red-winged blackbirds. Another useful approach would be further investigation of the Cuban red-winged blackbird, which is usually classified as a subspecies of *Agelaius phoeniceus*. Cuban redwings breed in marshes and are less sexually dimorphic in size, plumage, and behavior than are North American redwings (Whittingham et al. 1992a). For example, both sexes of the Cuban redwing sing the "conc-a-reeee" song, and both are black, although only males possess the red-and-yellow epaulet. Whittingham et al. (1992a) speculated that Cuban redwings are monogamous, but unfortunately, no direct observations of mating relationships have been published. Study of the mating system of Cuban redwings and of their phylogenetic relationship with North American redwings might shed much light on the origin of the mating system of the species as a whole.

Genetic Mating System

We classified the genetic mating system of red-winged blackbirds (see chapter 1) as polygynandry, a mating system in which both males and females share gametes with more than one member of the opposite sex within a single breeding season. The best evidence for a polygynandrous mating system comes from the DNA fingerprinting studies of Gibbs et al. (1990) and Westneat (1993a, pers. comm.). These studies showed that all young in every nest were the genetic offspring of the female attending the nest, and that overall about one-fourth of the young were sired by extrapair males, whereas the remaining young were sired by the owners of the territories on which they hatched. Between 40 and 50% of broods contained at least one chick resulting from extrapair copulation.

Both female and male behavior must contribute to producing the mix of extrapair and within-pair fertilization observed in red-winged blackbirds. Below, we consider female and male contributions in turn. In doing so, we first discuss possible costs and benefits of EPCs to each sex, and then behavior in the context of EPCs. We also discuss how males attempt to prevent extrapair fertilization (EPF) of their pair-bonded females.

FEMALE PARTICIPATION IN EPCs

Westneat et al. (1990) classified costs and benefits of EPCs to female birds as either phenotypic, affecting the female's own survival and reproduction, or genotypic, affecting the genotypic fitness of her young. As evidence of a phenotypic benefit of EPCs for female red-winged blackbirds, Westneat (1992a) showed that the number of fledglings produced per brood increased (though slightly) with the number of males sharing paternity in the brood (Kendall's tau = 0.22, N = 61 broods, P < 0.05). The mechanism producing this correlation was unclear. Males did not feed their extrapair offspring, at least not as nestlings, nor did females forage on the territories of their extrapair partners. Infertility of the eggs was very rare and was not associated with complete infertility of the male. It is possible, however, that extrapair males guarded their young on other territories. Consistent with this last possibility, Patrick J. Weatherhead and colleagues (pers. comm.) have found lower mortality among extrapair young of territorial males (who would have an opportunity to guard these young) than among extrapair young of floaters (who would have much less opportunity).

The most likely genotypic benefit of extrapair copulation is obtaining "good genes" (Birkhead and Møller 1992). Female redwings seem to choose where to settle largely on territory quality (chapter 5), so EPCs may offer a means of obtaining good genes when the territory owner is inferior genetically. The benefit of choosing a male for good genes is likely to be low, because of the low heritability of fitness, but nevertheless greater than zero (Williams 1975, Searcy 1982).

Of the various costs of EPC suggested by Westneat et al. (1990), the most serious would seem to be direct punishment by the pair-bonded male and withdrawal of male parental care. With regard to direct punishment, male redwings do attack their mates at times, but there is no evidence that these attacks are linked to EPCs, or that females are ever really injured. With regard to withdrawal of paternal care, pair-bonded males do not adjust their provisioning with respect to their paternity (chapter 2), but they may adjust their nest guarding. Patrick J. Weatherhead and colleagues (pers. comm.) found that territory owners defend nests less vigorously when the broods included one or more extrapair young (chapter 2). Moreover, in this study, broods with one or more young sired by EPC were significantly less likely to produce fledglings than were broods in which all young were sired by the territory owner, and the proportion of young sired by EPC was significantly lower among nestlings that survived than among those that did not.

In summary, the evidence is mixed on the balance of phenotypic costs and benefits of EPC to female red-winged blackbirds. In Westneat's study in New York, broods sired by multiple males produced more fledglings, but Weatherhead and colleagues found in Ontario that broods with extrapair young were more likely to fail. On balance, it seems likely to us that there is usually little or no net phenotypic benefit to females of pursuing EPCs. We would expect genetic benefits to be low also, on general principles. We would therefore predict that female redwings should be indifferent to EPCs. The most extensive study of courtship and copulation in red-winged black-birds is that of Westneat (1992a), done on a mixed marsh and upland popu-lation in New York. Westneat observed 410 instances of courtship, 86% of which were intrapair and 14% extrapair. Females crouched, indicating ac-ceptance of the male, in 29% of the intrapair courtships compared to only 7% of the extrapair courtships. Courtship never proceeded to copulation un-less the female crouched. Extrapair courtship was initiated by the male in all 18 of those instances that were adequately observed, i.e., the male ap-proached the female rather than vice versa, and the male was first to show courtship behavior. Females were never observed to enter a neighboring territory and solicit copulation from the owner. By contrast, females crouched before the male approached and courted in 54% of the 178 intra-pair courtships that were observed adequately. These data indicate that fe-males do not seek EPCs but rather seek intrapair copulations.

Westneat's data, however, contain a paradox: only 6% (4 of 71) of the observed copulations were extrapair, yet 23% of the offspring were sired by extrapair males. The other two studies of copulation in red-winged black-birds found somewhat higher frequencies of EPCs: 12% (seven of 58) in

Monnett et al. (1984) and 16% (nine of 56) in an unpublished study by Emily Davies, but again these frequencies are lower than the 23% extrapair paternity found by Westneat and the 28% found by Gibbs et al. (1990). One explanation of these puzzling results is that EPCs are more cryptic and thus more difficult to observe than within-pair copulations. Westneat does not credit this explanation; he asserts that females could be observed almost continuously when they were on territory and that his observations spanned all times of the day and season when copulations might occur. Females visited neighboring territories very rarely and could be observed throughout these visits. Females did at times make longer forays away from the neighborhood of their home territories, during which the observers often lost sight of them; however, the frequency of such forays per female was not associated with the frequency of EPFs in the broods (Westneat 1992a).

Another way to reconcile the frequencies of EPCs and EPFs is to hypothesize that the missing copulations occur during sexual chases (Westneat 1992a). In a sexual chase, one to several males pursue a rapidly fleeing female, often striking or grabbing her to end the pursuit (Nero 1956a). The male initiating the chase is most often the female's mate, but other males may participate as well (Nero 1956a). Sexual chases often end with the female on the ground, where a forced copulation might occur, but as far as we know forced copulations have never been observed in association with sexual chases in red-winged blackbirds (Westneat 1992a).

A third way to resolve the paradox is to suggest that EPCs are better timed than within-pair copulations relative to the female's fertile period. This seems unlikely, as the owner of the territory on which a female is nesting ought to have a better opportunity to observe her condition and the status of her nest than does any other male. Westneat (1993b) found that within-pair and extrapair courtships seemed about equally well timed relative to the presumed fertile periods of females.

In sum, female red-winged blackbirds have not been observed to pursue EPCs actively. We seem, however, to have missed observing some substantial fraction of EPCs, as the fraction of EPFs is higher than the observed fraction of EPCs. Females may or may not actively pursue EPCs during these missing encounters. Clearly, more research is needed on patterns of courtship and copulation.

MALE PARTICIPATION IN EPCs

We do not have to search for a fitness benefit of EPCs to a male: a male benefits directly by siring extra offspring. Balanced against this benefit are a number of possible (opportunity) costs, all having to do with the diversion of the male from other behaviors. Behaviors that may trade off against EPC include mate attraction, parental care, territory defense, and mate guarding.

Existing evidence indicates that males usually do not travel far in pursuit of EPCs. For example, of 13 EPFs demonstrated by Gibbs et al. (1990), eight occurred between a territory owner and a female nesting on an adjacent territory. In the five remaining cases, the male crossed from one to three intervening territories to reach the extrapair female. The distance traveled was never more than 200 meters. It may be that males do not travel far in pursuit of EPCs because they cannot monitor the reproductive condition of females on distant territories without diverting too much time from other activities. We suggest, then, that the frequency of EPCs is determined in part by a balance between the genetic benefits of EPCs and the opportunity cost of monitoring females on other territories, a cost that must rise sharply with distance. Another important factor must be that to participate in EPFs, a male must evade the countermeasures of the female's mate.

Male Countermeasures

Westneat (1993b) has investigated two types of countermeasures used by male birds in general to minimize EPFs: mate guarding and frequent within-pair copulation. He found that redwing pairs performed approximately 21 copulations per clutch, which indeed seems more frequent than would be needed without the threat of sperm competition. Males did not, however, increase within-pair courtship after another male had intruded onto the territory, nor did they increase courtship after female absences from the territory.

Male redwings followed females more than vice versa during the female's fertile period (Westneat 1993b), which is usually taken as an indication of mate guarding. Nevertheless, males spent only 33% of the time close to the female (<10 meters) (Westneat 1993b), less than in most instances of mate guarding (Birkhead and Møller 1992). Westneat believed, however, that males can guard their mates effectively by maintaining watch over their territories, without following individual females.

If keeping watch over the territory is important to mate guarding, we would expect males to increase their territory attendance during the fertile periods of resident females. Yasukawa (unpublished) tested this prediction for his Indiana study marsh. As shown in Figure 6.3, males spent 91 to 95% of their time on the territory just before the primary female's fertile period, and 90 to 97% just after, whereas during the female's fertile period, males increased time on territory to 97 to 100%. Westneat (1993b) found a similar pattern in his New York population. Westneat also found that EPCs were much more likely to occur when owners forayed off the territory than when they were on territory. When David F. Westneat (pers. comm.) removed territorial males and held them for one-hour periods, the frequency of EPCs increased dramatically: 36 successful EPCs were observed in 29 hours of

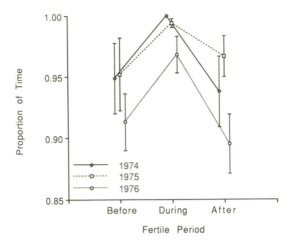

Figure 6.3 Proportion of time spent on territory by male owners in relation to the fertile period of the primary female, at Yellowwood Marsh in Indiana, from Yasukawa (unpublished). Plot shows mean ± SE for 23, 24, and 28 males observed in 1974, 1975, and 1976, respectively. Males spent a higher proportion of time on territory during the fertile period than before or after in each of three years.

removals, compared to none during 57 hours of control observations (P < 0.01).

Mate guarding should not conflict with territory defense, as the two behaviors require similar forms of vigilance and aggression, but it does seem to conflict with other behaviors, such as foraging. As noted above, male forays off the territory declined during mate-guarding periods in Westneat's (1993b) study, and Westneat believed that the purpose of most forays was to obtain food. When David F. Westneat (pers. comm.) provided supplemental food on territories, the frequency of male forays decreased significantly, whereas within-pair paternity increased significantly. Mate guarding may also trade off against attempts to attract new females to the territory. Westneat (1993a) found some evidence that within-pair paternity decreased when a new female settled during an established female's fertile period, but the effect did not occur in all analyses. Overall there was no relationship between harem size and within-pair paternity (Westneat 1993a).

Conclusions

In this chapter we have presented our short-term explanation of social polygyny and have speculated about long-term explanations as well. The short-term model that we favor for red-winged blackbirds has the following characteristics. First, it is a female-choice model, in which mating relationships are formed through female choice rather than through male coercion. Second, it is a no-cost, no-benefit model, in which females neither pay a net cost, nor receive a net benefit, as a result of polygynous mating. Third, our model assumes directed choice, in which females show preferences for certain territories and perhaps males, with the preferences being shared by different females, at least to some extent. We believe this explanation is well supported by the data on red-winged blackbirds.

We do not claim that this hypothesis is a general explanation of polygyny in birds; on the contrary, it is clear that different hypotheses must be invoked for other species (Searcy and Yasukawa 1989). What we do claim, however, is that there are general advantages to the method of analysis we have used in examining the redwing system. We refer here particularly to two aspects of the analysis: first, the use throughout of multiple, alternative hypotheses, and second, the use of removal experiments to investigate costs and benefits of polygyny to females.

We have reached no definite conclusion on the long-term explanation of polygyny in redwings. The major stumbling block is that any such explanation depends vitally on the assumed ancestral mating system, from which the present system originated, and we can only speculate as to what this ancestral system was like. We have suggested two, rather different scenarios, one assuming an ancestral system modeled on the standard mating system of monogamous passerines, and the other assuming an ancestral system resembling the mating systems of monogamous *Agelaius* species. Other ancestral systems are possible and would produce additional scenarios. The application of modern comparative methods should help to resolve some of these issues.

Study of the genetic mating system is relatively recent, so conclusions must be provisional. Observations to date indicate that females do not seek EPCs, instead showing some preference for within-pair copulations. This conclusion is especially provisional, as some substantial fraction of EPCs are evidently going unobserved. Males both actively seek EPCs with females on other territories and actively guard their own females against the EPC-seeking of other males. Mate guarding is accomplished through territory defense, and is quite effective in preventing EPFs, but is limited by the need of males to spend some time foraging away from their territories.

7   Sexual Selection in Progress

Darwin (1859, 1871) proposed that dimorphism in traits associated with display, ornamentation, and fighting have evolved due to sexual selection, in most cases acting on males. He also suggested that sexual selection is stronger in polygynous species, such as red-winged blackbirds, than in monogamous ones (see chapter 1). In this chapter and the next, we examine whether dimorphic traits can be explained by sexual selection acting on males in red-winged blackbirds. We employ two general sorts of tests, corresponding to Grafen's (1988) distinction between studies of selection in progress and studies of adaptation. As discussed in chapter 1, studies of selection in progress relate natural variation in a trait to variation in lifetime reproductive success or one of its components, and thus address the question of whether selection is currently acting on the trait. Studies of adaptation ask whether and how the trait is of use to the individual. We would expect that many adaptations would no longer be subject to selection, at least not directional selection, so studies of selection in progress may tell us little about adaptation. Grafen (1988) suggests that the best way to study adaptation will often be to manipulate the trait artificially and then make rather proximate measures of the effects of the manipulation, for example on female preferences or success in fighting. Other kinds of evidence are also appropriate in the study of adaptation, such as observations of patterns of use of the trait, and interspecific comparisons. These types of evidence, pertaining to adaptation but not to selection in progress, will be covered in the next chapter.

In the present chapter we review attempts to measure sexual selection in progress in red-winged blackbirds, that is, attempts to measure the current strength of sexual selection acting on various traits. This means relating natural variation in the trait to some measure of mating success. We begin by describing attempts to measure the "opportunity" for sexual selection, by which is meant the potential strength of this component of selection in the species. This will lead us to consider more carefully the components of sexual selection in red-winged blackbirds. We then describe the traits that are sexually dimorphic in redwings, and which are therefore candidates for the action of sexual selection. Next we will review methods used in relating natural variation in male traits to mating success and summarize the evidence with respect to each trait. For the most part, the evidence will bear on directional sexual selection, as opposed to stabilizing selection. We will also discuss the possibility that viability selection acts to counter sexual selection, and thus to stabilize sexually selected traits. Finally, we will try to explain why sexual selection in progress is not as strong as might be expected in red-winged blackbirds.

Opportunity for Sexual Selection

Sexual selection depends on variance in mating success. Therefore, it seems reasonable to expect sexual selection to be more powerful in nonmonogamous species, in which some individuals get several mates and others none, than in monogamous species, in which almost all adults obtain one and only one mate. Extrapair fertilizations can, however, produce variance in mating success in socially monogamous species, countering the expected trend. Testing the prediction requires that we make quantitative measurements of the power of sexual selection that can be compared between species differing in mating system. Unfortunately, the power of sexual selection to cause evolutionary change in any trait depends not only on the amount of variation in lifetime reproductive success that is due to variation in mating success, but also on the importance of the particular trait in determining individual mating success, and on the heritability of the trait (Clutton-Brock 1988). The latter two factors must necessarily vary between traits within a sex and species, and hence no single answer can be given as to the power of sexual selection in that sex and species. Therefore, attempts to make general comparisons of the strength of sexual selection between species are, in a way, doomed from the start. Nevertheless, many attempts have been made to do just that, usually by employing some measure of the first factor given above, the amount of variation in lifetime reproductive success due to variation in mating success. These measurements do not gauge the actual strength of sexual selection, but rather its potential strength, so they have recently been termed measurements of the "opportunity" for sexual selection (Clutton-Brock 1988).

Payne (1984) and Clutton-Brock (1988) together have listed nine indices of the opportunity for sexual selection, some reasonable and some rather useless. One of the more reasonable indices has been applied to red-winged blackbirds: the ratio of the variance in breeding success of males to the variance in breeding success of females. The original estimates were made by Payne (1979) for a population in a marshy, streamside habitat in Michigan. Payne banded all the territorial males and about half the breeding females. Banded females were matched to nests by observation during incubation and feeding of young. Total numbers of females per territory were estimated as the sum of the banded females, unbanded females associated with a nest, and unbanded females not associated with a nest. The last category made up less than 10% of the whole. Reproductive success of females was estimated as the number of young fledged from their nests, and for males as the number of fledglings produced on their territories. Thus, Payne assumed that males sired all the young raised on their territories.

Payne made three different assumptions about which classes of males to include in "the male population": (1) territorial males only, (2) territorial males plus adult males captured on the study site but not holding territory there, and (3) territorial males, nonterritorial adults, plus one-year-old males captured on the study site (none of which held territory). As shown in Table 7.1, estimates of the variance of male reproductive success are substantially lower with the third assumption because of the addition of large numbers of one-year-old males all having zero success. Payne argued, with reason, that at least some of the one-year-old males should be excluded, on the grounds that they occupy much larger areas than the study site. This may be true also of the nonterritorial adult males. At any rate, with any of our assumptions about male population size, variance in breeding success among males does tend to be greater than among females; the differences in variances between the sexes are significant in one year (1977) but not the other (1975).

We can also calculate from Payne's (1979) data I_m, an index of the overall "intensity of selection" on males (Table 7.1). I_m is the standardized variance in offspring number, i.e., the variance in offspring number divided by the square of the mean (Crow 1958). Wade and Arnold (1980) have shown that (with certain assumptions): $I_m = RI_f + I_s$, where R is the sex ratio (males/females), I_f is the standardized variance in offspring number for females, and I_s is the intensity of sexual selection on males (see also Wade 1979). Using Payne's data on I_m and R for all males, we find that I_s is 2.85 for 1975 and 4.86 for 1977, and that the ratios I_s/I_m are 0.60 and 0.78, respectively. The estimates of I_m leave out effects of variation in between-year survival, and this omission exaggerates the relative importance of I_s. The calculations at any rate must be very approximate, but they do make the point that the intensity of sexual selection on males is a substantial proportion of the overall intensity of selection.

Orians and Beletsky (1989) have also measured the relative variance in breeding success of the two sexes for a population of red-winged blackbirds, this one occupying marshes in eastern Washington State. Theirs was a large-scale, long-term study, in which an area containing 70 to 80 territories was followed for 10 years. As a result of the long duration of the study, Orians and Beletsky were able to estimate lifetime reproductive success, rather than just seasonal success as Payne (1979) did. Not all breeding females were banded, and some nests could not be associated with any female; therefore data on breeding success of females are incomplete. In some cases gaps

Table 7.1

Variance in Reproductive Success and the "Intensity of Selection" (I_m) on Male Red-winged Blackbirds.

Reproductive Success Measure	Year	Sample	N	Variance	I_m	Reference
Apparent Seasonal Success	1975	F (all)	28	3.01	—	Payne 1979
		M (territorial)	15	4.98	0.41	
		M (adults)	29	5.59	1.74	
		M (all)	61	3.43	4.75	
	1977	F (all)	40	2.52	—	
		M (territorial)	15	11.17	0.41	
		M (adults)	33	11.80	2.12	
		M (all)	77	6.36	6.23	
Apparent Lifetime Success	1977–	F (breeding)	795	13.13	—	Orians & Beletsky 1989
	1986	M (territorial)	264	195.65	—	
Apparent Seasonal Success	1986	M (territorial)	13	9.47	0.25	Gibbs et al. 1990
Actual Seasonal Success	1986	M (territorial)	13	11.23	0.40	
Apparent Seasonal Success	1988	M (territorial)	26	10.7	1.04	Westneat 1993a
Actual Seasonal Success		M (territorial)	25	9.6	1.32	
Apparent Seasonal Success	1989	M (territorial)	25	10.2	1.00	
Actual Seasonal Success		M (territorial)	24	9.8	1.09	

were obvious, i.e., when females were known to breed in two or more years but were not located in an intervening year. In these cases, Orians and Beletsky credited the female with the average number of fledglings produced by experienced females in that year. All territorial males were banded. The measure of breeding success was again the number of fledged young, and again it was assumed that males sired all the young produced on their territories.

Among the 264 males known to have bred, variance in lifetime reproductive success was 195.7, compared to a variance of 13.1 for 795 females. The ratio of male to female variance, then, was 14.9. This variance ratio is considerably larger than in Payne's study; the difference is no doubt in part due to Orians and Beletsky's (1989) considering lifetime success whereas Payne (1979) used seasonal success, but the larger mean and maximum harem sizes in Orians and Beletsky's (1989) western study area must also have played a role. Orians and Beletsky (1989) showed using multiple regression that 39.8% of the variance in lifetime reproductive success among males was explained by number of years breeding, 31.5% by the mean fledging success of females, and 23.6% by harem size. The last figure could be taken as another measure of the opportunity for sexual selection among male red-winged blackbirds.

There is, however, a major problem with both Orians and Beletsky's (1989) and Payne's (1979) data, in that we know that it is not always safe to assume that the territory owner sires all the young produced on his territory. The best data on actual paternity come from DNA fingerprinting studies, which are capable not only of excluding individuals who are not true parents but also of identifying the true parents with some confidence. Gibbs et al. (1990) used this technique to determine parentage of redwing young on three marshes in Ontario (Figure 7.1). For 13 territories on one marsh, Gibbs et al. (1990) estimated the actual reproductive success of the territory owners as the sum of the young sired on territory and off territory. In addition, they estimated "apparent reproductive success" in the conventional way, as the number of young produced on territory. Variance in actual success was, if anything, greater than variance in apparent success (Table 7.1) and was as great or greater than the variance estimates (for all territorial males) in Payne's (1979) study, which also looked at seasonal reproductive success only. The intensity of selection (I) was also somewhat greater when estimated from actual success than when estimated from apparent success.

A second DNA fingerprinting study was carried out by Westneat (1993a) in a mixed marsh and upland site in New York. In estimating actual reproductive success, Westneat was able to assign paternity to nearby territory owners for 33 of the 55 young for which within-pair paternity was excluded. Variance in apparent reproductive success of males was slightly higher than variance in actual success in each of two years (Table 7.1), but the differ-

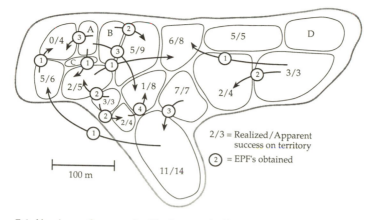

Figure 7.1 Numbers of young sired both on and off territory by male red-winged black-birds in an Ontario marsh, from Gibbs et al. (1990). Putative parents were excluded by first screening families for cases in which an MHC (major histocompatibility complex) probe revealed a mismatch between parents and offspring at a single hypervariable locus, and then confirming the mismatch using the complex banding pattern detected by one or two multilocus minisatellite probes. In cases in which a putative parent was excluded, the actual parent was identified by a similar two-step procedure. The frac-tions within each territory give the number of young sired on the territory by the owner divided by the total number of young hatched on the territory. Arrows indicate cases in which a territory owner sired young on another territory; each arrow originates on the territory of a male responsible for siring extrapair offspring, the arrow points to the territory on which the extrapair offspring were hatched, and the circled number gives the number of young involved. Reprinted with permission from *Science*, December 7, 1990. Copyright AAAS.

ences were quite small. Estimates of I were also slightly higher for actual than for apparent success.

Overall, the estimates of variance in seasonal reproductive success for territorial males from Payne (1979), Gibbs et al. (1990), and Westneat (1993a) are all remarkably similar (Table 7.1). This gives us some justifica-tion for continuing to use Orians and Beletsky's estimates of variance in lifetime reproductive success, even though these estimates are based on ap-parent rather than actual success. Table 7.2 compares Orians and Beletsky's (1989) estimate of the ratio of male variance to female variance in lifetime reproductive success to estimates for other species of birds. It can been seen that the ratio in redwings is far higher than in any of the other species. The other eight species in the table are classified as socially monogamous, with the exception of pied flycatchers. Pied flycatchers are termed polygynous, but they exhibit a much lower degree of polygyny than do redwings (Table 1.2) and perhaps should be termed "facultatively bigynous." In either case, pied flycatchers show a ratio of male-to-female variance that is in the middle of the range for monogamous species. Table 7.2 also shows the maximum

Table 7.2
Variance in Lifetime Reproductive Success of the Sexes in Various Bird Species.

Species	Sex	N	Maximum	Variance	Male/Female	Reference
Red-winged Blackbird	Males	264	159	195.6	14.9	Orians & Beletsky 1989
	Females	795	24	13.1		
Pied Flycatcher	Males	953	37	32.6	1.0	Sternberg 1989
	Females	1298	36	32.9		
House Martin	Males	103	42	54.8	2.6	Bryant 1989
	Females	125	28	21.2		
Kingfisher	Males	74	26	37.1	1.0	Bunzel & Drüke 1989
	Females	51	32	38.3		
Meadow Pipit	Males	49	25	35.6	2.6	Hötker 1989
	Females	33	12	13.9		
Splendid Fairy-wren	Males	36	—	90.3	1.0	Rowley & Russell 1989
	Females	46	—	88.4		
Osprey	Males	33	18	42.4	0.9	Postupalsky 1989
	Females	51	29	49.3		
Mute Swan	Males	58	37	54.3	0.8	Bacon & Andersen-Harild 1989
	Females	70	37	64.7		
Red-billed Gull	Males	81	6	4.7	0.5	Mills 1989
	Females	66	9	8.9		

NOTE: LRS measured as number of fledglings produced.

number of young produced per lifetime by a male and a female within each species. Note that the 159 offspring produced by a male redwing is far greater than for any individual in any of the monogamous species, and that the ratio of male maximum to female maximum is likewise greater for redwings than for the other species.

The data on ratios of variance in lifetime success fit well with the intuitive idea that sexual selection ought to be stronger in the highly polygynous red-winged blackbird than in most birds. Given the problems inherent in the basic concept of measuring the overall strength of sexual selection, the debate on what measures to use (Clutton-Brock 1988), and the uncertainties of estimating variance in reproductive success, perhaps not too much should be made of this apparent fit.

Potential Rates of Reproduction

A recent entry in the derby of ways to estimate the intensity of sexual selection is based on two factors: relative parental investment (PI) and the operational sex ratio (OSR). OSR is defined as the relative availability of males and females that are ready to mate. Both PI (Trivers 1972) and OSR (Emlen and Oring 1977) have been proposed as critical factors determining competition for mates, and therefore sexual selection, but each has its share of problems. PI is probably impossible to measure in species in which both sexes contribute parental care (chapter 5), and it does not predict the pattern of mating competition in such species (Clutton-Brock and Vincent 1991). OSR probably does affect the intensity of competition for mates, and therefore the intensity of sexual selection, but many factors affect OSR, and it cannot be measured simply as the adult sex ratio (Clutton-Brock and Parker 1992). Clutton-Brock and Parker (1992) have argued, however, that potential rates of reproduction (PRR; see chapter 5) of males and females are closely related to OSR, are constrained by physiological and environmental factors, and are affected by the relative expenditures of male and female parents on their progeny.

According to this view, PRR for males and females can be used to estimate the OSR of a species and thereby give a clear indication of the relative intensity of sexual selection on the two sexes. As a first approximation, we estimated PRR in red-winged blackbirds as 12 fledglings/year for females and 180 fledglings/year for males in western populations (chapter 5). According to these estimates, male PRR is 15 times that of females, indicating an OSR strongly biased in favor of males. Similar calculations using estimates from eastern populations with smaller maximum harem sizes still suggest an OSR that is male biased, though less strongly (chapter 5). A male-biased OSR predicts competition among males for access to females (Clutton-Brock and Vincent 1991, Clutton-Brock and Parker 1992).

Components of Sexual Selection

We will not attempt to relate variation in male traits to variation in lifetime reproductive success, in part for the obvious reason that we do not have accurate data on lifetime reproductive success in males, and in part for the less obvious reason that lifetime reproductive success measures natural selection as well as sexual selection. Our principal interest is in sexual selection, so we will concentrate on relating male traits to measures of success under sexual selection, that is, to measures of male mating success.

We will consider variation in mating success among red-winged blackbirds to have four main components: (1) a difference in mating success between males that own territories (owners) and males that do not (floaters); (2) differences in mating success among owners due to differences in harem sizes; (3) differential success of owners in copulating with females from their own harems; and (4) differential success of owners in extrapair copulation. One reason for dividing things up this way is practical: it allows us to separate the components of mating success that have been related to male traits (1 and 2) from those that have not yet been investigated (3 and 4). Components 1 and 2 arguably stem from intrasexual competition for territory, 1 for obvious reasons, and 2 because differences in harem sizes seem to be due primarily to female choice on territory quality (chapter 5). Components 3 and 4 presumably stem directly from intersexual selection.

These components are all measures of mating success, but each measures selection only if it is correlated with reproductive success. Previously it was thought that harem size was strongly correlated with male reproductive success, because strong correlations had been demonstrated between harem size and the number of young raised per territory (Searcy and Yasukawa 1983, Orians and Beletsky 1989). DNA fingerprinting, however, has since demonstrated the frequent occurrence of extrapair copulations, which weaken the relationship between harem size and male reproductive success by introducing two sources of variation in reproductive success unrelated to harem size: variation in within-pair paternity and in extrapair fertilization (Gibbs et al. 1990, Westneat 1993a). Taking these factors into account, the relationship between harem size and male reproductive success appears to be positive but fairly weak, with nonparametric correlation coefficients in the range of 0.13 to 0.51 (see chapter 4).

The discovery that extrapair fertilizations are common in red-winged blackbirds has not changed the assumption that territory owners have much higher mating and reproductive success than floaters, because floaters apparently participate little in extrapair copulations. Three lines of evidence point to this conclusion. First, Bray et al. (1975) found that the proportion of infertile eggs on the territories of vasectomized males increased when surrounding territory owners were also vasectomized, implying that these other territory owners were performing most of the extrapair copulations. Second, in studies in which copulations were directly observed, females were seen to copulate only with the owner of their own territory or of a neighboring territory (Monnett et al. 1984, E. Davies pers. comm.). Third, the DNA fingerprinting studies have assigned paternity to other territorial males in most cases of extrapair fertilization. Gibbs et al. (1990) found that 26 of 28 offspring resulting from extrapair fertilization were sired by other territory owners. Westneat (1993a) also found that the majority of young resulting from extrapair fertilizations could be assigned to nearby territory owners; most of the exceptions were young hatched near the edge of the study area, where there were nearby territorial males who had not been fingerprinted.

More recently, Patrick J. Weatherhead and colleagues (pers. comm.) found in an extensive fingerprinting study that one in five extrapair offspring could not be assigned to a territory-owning father. Since nearly all owners were sampled in this study, Weatherhead and colleagues concluded that these young must have been sired by floaters. Assuming that floaters do sire about 20% of the extrapair offspring, and remembering that about 25% of offspring result from extrapair fertilizations, it follows that floaters may have sired about 5% of the total young. Considering that floaters probably out-number owners substantially (chapter 3), these estimates confirm that the average reproductive success of floaters must be very low indeed.

A male who fails to obtain a territory in a given year must have a good chance of dying without ever obtaining a territory, and thus without ever achieving any substantial mating success. To see this, we can use data from Orians and Beletsky's (1989) long-term study on ages at which male red-winged blackbirds first held territories for more than a small fraction of the breeding season. Of the sample of 148 males who held territory at some time, 11 (7%) first obtained a territory at one year of age, 84 (57%) at two years, 34 (23%) at three years, 16 (11%) at four years, one (0.7%) at five years, and two (1.4%) at six years. If we assume that the survival rate of males is a constant 0.54 per year from one year of age on (Searcy and Yasukawa 1981a), and that no males survived past six years without holding territory, then we can use these figures to reconstruct a cohort of one-year-olds from which these first-time territory owners are drawn (Box 7.1). The original size of the cohort is 441 one-year-old males; 148 of these eventually obtain territories, leaving 293 (66%) to die before obtaining a territory. Of the 232 two-year-olds, 95 (41%) die before obtaining a territory. This recon-struction must be only approximate, but it is nonetheless clear that mortality rates are high enough that a male failing to obtain a territory in a given year stands a substantial risk of dying before achieving any reproductive success. This indicates that there must be a strong selective advantage to rapid suc-cess in obtaining a territory.

BOX 7.1 RECONSTRUCTION OF A
COHORT OF TERRITORIAL MALE
RED-WINGED BLACKBIRDS

Data are given by Orians and Beletsky (1989) on the numbers of males observed to obtain their initial territories as one-year-olds, two-year-olds, three-year-olds, etc., for a population in Washington State. From these data, we can estimate the number of males at each age competing for an initial territory, i.e., the number of males of each age class that are alive at the start of the breeding season and that have not yet held territory. We do this by reconstructing the cohort assuming a constant probability of survival of 0.54 per year from the age of one onward, as follows.

We arbitrarily assume that no males without territories survive beyond age six, the oldest age at which males were observed to obtain initial territories. Then the two males that had not yet held territory at the start of age six should represent 2/0.54 = four males that had not yet held territory at age five. One male obtained an initial territory at age five, giving five total five-year-olds that were alive at the start of the breeding season and had not yet held territory. These five males should represent 5/0.54 = nine males that had not yet held territory at age four. Sixteen males obtained an initial territory at age five, so 16 + 9 = 25 four-year-old males that were alive at the start of that breeding season and had not yet held territory. By reiterating this logic, we work backward to reconstruct the entire cohort, as shown below.

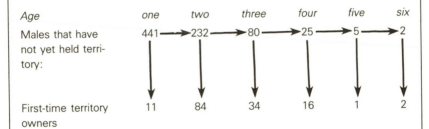

Age	one	two	three	four	five	six
Males that have not yet held territory:	441 →	232 →	80 →	25 →	5 →	2
First-time territory owners	11	84	34	16	1	2

According to this reconstruction, then, the proportion of one-year-olds obtaining an initial territory is 11/441 = 0.024. The proportion of two-year-olds that have not yet held territory who succeed in obtaining one is 84/232 = 0.362; the proportion of three-year-olds is 34/80 = 0.425; and the proportion of four-year-olds is 16/25 = 0.640. Beyond age four the numbers are too small to conclude much anyway, but note that 100% of six-year-olds obtain territories by assumption rather than by result.

Presumably, the success of owners in copulating with their own and other females (components 3 and 4) should correlate with reproductive success. Unfortunately, no data have yet been published relating such measures of copulation success to male traits in red-winged blackbirds, though such data are now being gathered (Patrick J. Weatherhead pers. comm.). We look forward to the publication of these data with keen anticipation.

Sexually Dimorphic Traits

The existence of sexual dimorphism in a trait is itself preliminary evidence that the trait has been acted on by sexual selection; however, it is only weak evidence as there are other selective forces besides sexual selection that can cause sexual dimorphism (Darwin 1871). In red-winged blackbirds, sexual dimorphism occurs principally in traits associated with: (1) visual display, (2) vocal display, (3) body size, and (4) aggressiveness. Below, we describe sexual dimorphism in each of these categories and discuss problems encountered in quantifying natural variation in the traits within each category.

VISUAL DISPLAY

The principal visual display trait of male red-winged blackbirds is the brilliant red epaulet, with its yellow border. Some females, usually older ones, also have epaulets, but these are always much smaller than in males, and the color is duller, ranging from pale yellow-orange to red-orange, rather than from red-orange to red as in adult males. Yearling males have epaulets intermediate between those of females and older males; colors range from orange to red-orange, and the epaulet is often flecked with brown or black.

Mating success has been related to variation in both size and color of epaulets. Epaulet size can be measured as the distance from the bend of the wing to the farthest extent of red plumage ("length of red") or to the farthest extent of yellow plumage ("length of yellow"). Color can be measured as steps on a continuum from orange to red. For example, male epaulet colors match nicely to the Royal Horticultural Society's six-step scale from orange (12) to vermilion red (18).

VOCAL DISPLAY

The principal vocal display in males is the song, which we described in chapter 3. Females also have songs (see chapter 9), but these are entirely distinct from male song. Studies of sexual selection have concentrated on two measures of natural variation in male song: song rate (e.g., songs per minute) and song repertoire size (the number of distinct song types in a male's repertoire). Males give two other vocalizations in courtship and territory defense, growls and precopulatory notes; we will discuss these briefly in chapter 8.

BODY SIZE

It is obvious to any observer than male redwings are considerably larger than females. Quantifying this size dimorphism and the natural variation in size among males, however, is complicated by the difficulty of defining body size. In some ways, the problems of defining body size in birds are similar to those of defining intelligence in humans: it is clear that certain measures are related to body size (or intelligence), but whether any one measure or even combination of measures *is* body size (or intelligence) is arguable. Among the univariate measures used to represent body size in birds, the most popular have been mass, wing length, and tarsus length. Additionally, many studies have used multivariate measures, produced by taking a variety of external measurements (such as mass, wing length, and tarsus length) and/or internal measurements of bones, reducing them to a few linear combinations using principal component analysis, and assuming that the first principal component represents size. None of these alternatives are perfect. Mass of an individual can vary within a day or over a season, and therefore mass tends to show the greatest variance of any of the popular size measures (Rising and Somers 1989). Wing lengths also tend to vary seasonally due to feather wear, and can also increase from year to year (following molts) even within adults. Tarsus lengths (and lengths of other single bones) do not change within adults, but they are not very intuitively satisfying, perhaps because they seem to relate to body shape as much as to body size. First principal component scores are rather abstract and have the additional drawback of being poorly standardized, i.e., the univariate measures on which they are based, and the statistical methods used to produce them, tend to vary from study to study.

If we were to recommend a body-size measure for future use, we would follow recent critiques (Rising and Somers 1989, Freeman and Jackson 1990) in recommending a principal component measure. Here, however, we will be reviewing studies which have variously used all of the options discussed above. Fortunately, the different measures do tend to be at least somewhat intercorrelated. Table 7.3 shows correlations between various body-size measures of adult male and female red-winged blackbirds, from a sample collected by Freeman and Jackson (1990) in a single county in Washington State. "PC1-bones" is the first principal component from a reduction of 13 skeletal measurements, whereas "PC1-flats" is the first principal component from a reduction of four external measurements (mass, and wing, tarsus, and tail lengths). Note that for males the univariate and multi-

Table 7.3
Correlations between Various Measurements of Body Size in Red-winged Blackbirds.

	Mass	Wing Length	Tail Length	Tarsus Length	PC1-bones	PC1-flats
Mass	—	0.16	0.11	0.15	0.28	NA
Wing Length	0.46**	—	0.73**	0.16	0.35*	NA
Tail Length	0.47**	0.67**	—	0.18	0.41*	NA
Tarsus Length	0.46**	0.37**	0.43**	—	0.48**	NA
PC1-bones	0.50**	0.48**	0.37**	0.72**	—	0.44**
PC1-flats	NA	NA	NA	NA	0.69**	—

*P < 0.05.
**P < 0.01.
SOURCE: Freeman and Jackson (1990).
NOTE: Correlations for adult males are below the diagonal, for females above. PC1-bones and PC1-flats denote first principal components from analyses of skeletal and external measurements, respectively. NA ("not applicable") indicates correlations between univariate measures and multivariate measures that contain them. Sample sizes are 53–60 for males and 36–46 for females.

variate measures are all strongly intercorrelated, whereas for females only PC1-bones shows consistently good correlations with other measures. Thus, for male redwings there is some justification for using the various measures interchangeably, whereas there is rather less justification for doing so for females. At any rate, we will continue to refer to a variety of these measures as "body size," but we will always state what specific measure is actually used to represent size in a given study.

Adult male redwings are about 10% larger than females in tarsus length, 20% larger in wing length, and 50% larger in mass. A similar index of dimorphism cannot be given for principal component measures because such measures are constructed so as to have a mean of zero.

AGGRESSIVENESS

Male red-winged blackbirds are certainly more aggressive than females, engaging in more frequent and more violent aggressive acts. Quantifying dimorphism is again difficult, however. One way to measure aggressiveness is to quantify aggressive response to a standard stimulus. For example, Searcy (1979b) used the latency of male territory owners to attack a caged male placed on their territories, whereas Rohwer (1982) used a principal component measure combining a number of aggressive behaviors shown by territory owners toward stuffed mounts. Because the same stimulus cannot be used to elicit intraspecific aggression from both males and females, one cannot generate a strictly comparable measure for both sexes. It is easier to compare the sexes in terms of interspecific aggressiveness; for example, Knight and Temple (1988) found that males made seven times as many attacks per time as females on models of American crows (chapter 2). We do not know, however, whether it is safe to assume that variance in interspecific aggression parallels variance in intraspecific aggression.

A male trait that is definitely related to aggressiveness, and which can be quantified, is the level of testosterone in the blood plasma. Plasma testosterone levels appear to be an order of magnitude lower in female redwings than in males (Beletsky et al. 1989, pers. comm.; Johnsen 1991) Testosterone levels are high in males during periods of the breeding season when aggression is common, both in red-winged blackbirds (Beletsky et al. 1989) and other birds (Wingfield et al. 1987). Furthermore, testosterone has been shown to stimulate aggression in males in redwings (Searcy and Wingfield 1980) and other species (Wingfield et al. 1987). It is also known, however, that aggressive interactions can lead to higher testosterone levels (Wingfield 1988). In other words, testosterone can stimulate aggression, but aggression can stimulate testosterone as well. Therefore, studies relating natural variation in testosterone levels to mating success must be interpreted cautiously.

Male Traits versus Mating Success

METHODS

One approach used to measure sexual selection in progress has been to compare traits of winners and losers in competition for territory; this focuses attention on our mating success component 1. A number of studies have taken this approach, using somewhat different methods.

Yasukawa (1979) compared males who were successful in establishing territories for the first time to unsuccessful males in terms of wing lengths and aggressiveness. Successful males were ones who held territories in breeding habitat during the period when females were nesting, regardless of whether they actually acquired females on their territories. Unsuccessful males held territories only before or after nesting, or held territories in habitat unsuitable for breeding. Weatherhead et al. (1987) compared sizes of males captured in a spring roost (representing the general population) with sizes of territory owners.

Eckert and Weatherhead (1987a) removed territory owners and then caught and measured replacement males; here the original owners represent the winners, and the replacement males (who only obtained a territory through the intervention of the experimenters) represent the losers. Owners and replacement males were compared for body size and epaulet size. Eckert and Weatherhead (1987a) also compared males that were displaced from their territories to males that successfully retained their territories through the season; here the displaced males represent losers and the "static" males represent winners.

Beletsky et al. (1989) compared testosterone levels in males captured as owners and males captured as floaters. Shutler and Weatherhead (1991a, b) removed owners to obtain replacement males and then removed successive replacements. First replacement males on a given territory were termed "shallow floaters" and all later replacements "deep floaters." Owners, shallow floaters, and deep floaters were compared on a number of morphological measures, including two measures of epaulet size and three measures of body size, and on mean song rates adjusted for time-of-day and season effects.

A second approach used to measure sexual selection in progress has been to correlate traits of territory owners with their harem sizes. This focuses attention on our mating success component 2. For example, Searcy (1979b) measured harem sizes and male traits for a population of males in Washington during each of three years. He then calculated correlations for each year between harem sizes and male traits such as epaulet size and color, song rates, wing length, mass, and aggressiveness (latency to attack a caged male). Other studies have similarly related harem sizes to characteristics of territory owners such as (1) song repertoire sizes (Yasukawa et al. 1980), (2) wing lengths (Yasukawa 1981a), (3) plasma testosterone levels (Beletsky et al. 1989), and (4) adjusted song rates (Shutler and Weatherhead 1991b). Eckert and Weatherhead (1987c) compared wing lengths in males holding territories in marsh and uplands in Ontario. Males in marshes have higher average harem sizes in this area.

Male age, or years of breeding experience, is a possible confounding factor in studies of sexual selection in progress in red-winged blackbirds. Yearling males have lower mating success than adults and also differ from adults in size, visual display, and other traits (see below). In this section, we consider only comparisons made either within adults or within yearlings. Within adults, Yasukawa (1981a) found a strong correlation between harem size and years of breeding experience in an Indiana population ($r_s = 0.706$, N = 54, P < 0.01). Similarly, Picman (1980b) found harem sizes of individual males increased between years in British Columbia. Orians and Beletsky (1989), however, found no tendency for individuals to increase their harem sizes between their first and second years of breeding in a large sample of Washington males (N = 148). Also in Washington, Searcy (1979b) found that experienced males (ones that had held territory previously) had larger harems than inexperienced males in one year but not in another. The only studies that have corrected for effects of age and experience are those of Yasukawa et al. (1980) and Yasukawa (1981a) on the Indiana population where experience effects seem most pronounced.

EPAULET SIZE AND COLOR

Results on sexual selection in progress on epaulet size and color are summarized in Table 7.4. The evidence indicates that epaulet size does not correlate with harem sizes among males, nor does epaulet size differ between winners and losers in competition for territory. Altogether there are 18 tests for an advantage of males with larger or brighter epaulets, only one of which gives a significant result at the 0.05 level; this corresponds very closely to chance expectations. We conclude that sexual selection in progress does not favor larger or redder epaulets.

SONG RATES AND SONG REPERTOIRE SIZES

We do not summarize the data on selection in progress on song in a table, as much of the evidence has not been reported in an appropriate form. Searcy (1979b) measured song rates for five different periods in three years, with samples of 17 to 36 males per period. Number of songs per observation period gave significant, positive correlations ($P < 0.05$) with harem size in two of the five periods; the correlation in another period was of equal magnitude and negative. Number of songs per minute on the territory gave significant correlations for none of the periods. Yasukawa (1979) found that 17 males that failed to establish territories had song rates equal to those of 37 successful males. Shutler and Weatherhead (1991b) found no significant relationships between adjusted song rates and harem sizes in either of two years for sample sizes of 79 and 57 males. Furthermore, they found no significant differences in adjusted song rates between owners, shallow floaters, and deep floaters for samples of 126 and 105 males. We conclude that there is little evidence of sexual selection in progress favoring higher song rates.

Yasukawa et al. (1980) found a significant positive correlation between repertoire size and harem size ($r_s = 0.396$, $N = 49$, $P < 0.01$). As mentioned above, however, there was a strong tendency for harem sizes to increase with male experience in this population. Repertoire size also tended to increase with increasing years of breeding experience ($r_s = 0.483$, $P < 0.01$). The partial correlation between repertoire size and harem size with years of breeding experience held constant was quite small (Kendall's partial tau = 0.102). When analysis was restricted to inexperienced males, the correlation between repertoire size and harem size was near zero ($r_s = -0.071$, $N = 28$, $P > 0.05$). It seems, then, that the relationship between repertoire size and harem size is mainly indirect; males with large repertoires have large harems, at least for the most part, because they also have more years of breeding experience. Again, there is no convincing evidence that sexual selection in progress favors larger repertoires.

Table 7.4

Sexual Selection in Progress on Epaulet Size and Color in Male Red-winged Blackbirds.

Evidence				Reference
Correlations of Harem Size in:	1974 (42)	1975 (34)	1976 (41)	Searcy 1979b
Length of Yellow	0.112	−0.092	0.194	
Length of Red	−0.153	0.197	0.122	
Color	0.084	−0.308	0.263*	
Comparisons of:	Territory Owners		Floaters	Eckert & Weatherhead 1987a
Length of Yellow (± SE)	44.0 ± 0.6 (23)		43.9 ± 0.4 (31)	
Comparisons of:	Static Males		Replaced Males	
Length of Yellow (± SE)	44.0 ± 0.3 (69)		43.7 ± 0.5 (35)	
Comparisons of:	Marsh		Upland	Eckert & Weatherhead 1987c
Length of Yellow (± SE)	43.5 ± 0.7 (29)		43.8 ± 0.6 (29)	
Comparisons of:	Territory Owners	Shallow Floaters	Deep Floaters	Shutler & Weatherhead 1991a
Length of Red (± SD)	37.7 ± 1.9 (56)	37.9 ± 1.7 (23)	37.1 ± 1.9 (35)	
Length of Yellow (± SD)	43.6 ± 2.9 (56)	43.7 ± 1.9 (23)	43.0 ± 2.7 (35)	

NOTE: Sample sizes are in parentheses.
*P < 0.05.

Body Size

Table 7.5 summarizes the evidence on sexual selection in progress on male size. We concentrate on the results on wing length since many studies used only this measure. A total of 15 relevant comparisons are made using wing length. Of these, only four are significant at the 0.05 level or better; three of these show a large male advantage and one a small male advantage. Ignoring the statistical significance of the individual comparisons, 11 are in the direction showing large male advantage, and four are in the opposite direction. The trend toward an excess of results showing large male advantage is itself not quite statistically significant (P = 0.06 by a sign test).

A possible criticism of the results on wing length and male success is that wing lengths tend to increase slightly with age, and therefore any tendency for success to increase with age will tend to produce positive associations between size and success. The results from Searcy (1979b) on mass and from Shutler and Weatherhead (1991a) on mass and tarsus length, however, tend to show the same trends as with wing length: positive relationships between size and success which are weak and mainly nonsignificant. Furthermore, Yasukawa (1981a) found a positive correlation between wing length and harem sizes among a sample of males all of whom were inexperienced ($r_s = 0.414$, N = 39, P < 0.01).

We conclude that there is some evidence of sexual selection in favor of large size in males, but such selection appears to be weak, at best.

Aggressiveness and Testosterone Levels

Again, much of the evidence on aggressiveness and testosterone levels cannot be summarized in tabular form. Yasukawa (1979) found that successful males, if anything, had display rates indicating they were less aggressive than unsuccessful males. Searcy (1979b) found no significant correlations between harem sizes and latency to attack a decoy in samples of 17 males in each of two years. Rohwer (1982) found that owners were more aggressive toward taxidermic mounts than were their replacements on 15 of 20 occasions; shallow floaters were more aggressive than deep floaters on 9 of 21 occasions. Thus, the evidence on behavioral measures of aggressiveness is mixed.

Beletsky et al. (1992) found a weak but significant correlation ($r_s = 0.18$, P < 0.05) between harem size and early-April plasma testosterone levels of 113 males. In another study, the same authors found that territory owners had significantly higher testosterone levels (P < 0.05) than floaters in two of three monthly comparisons (Beletsky et al. 1989). Again, interpretation of these results is difficult because of the possibility that territory ownership

Table 7.5

Sexual Selection on Male Body Size.

Correlations with Harem Size in: (Reference: Searcy 1979b)

Evidence	1974 (42)	1975 (34)	1976 (41)
Mass	0.300*	0.116	0.215
Wing Length	−0.210	0.073	0.319*

Comparison of Males: (Reference: Yasukawa 1979)

	Obtaining Territory	Not Obtaining Territory
Wing Length	126.7 (37)	126.0 (17)

Correlation between Harem Size and Male Wing Length: 0.373** (54) (Reference: Yasukawa 1981a)

Comparison of: (Reference: Eckert & Weatherhead 1987a)

	Territory Owners	Nonowners
Wing Length (± SE)	124.6 ± 0.6 (24)	122.9 ± 0.7 (32)

	Static Owners	Replaced Owners	Displaced Owners
Wing Length (± SE)	123.9 ± 0.4 (71)	123.3 ± 0.5 (21)	124.0 ± 0.6 (8)

Comparison of: (Reference: Eckert & Weatherhead 1987c)

	Marsh	Upland
Wing Length (± SE)	119.5 ± 0.9 (29)	122.3* (± 0.9) (29)

Comparison of: (Reference: Weatherhead et al. 1987)

	Territory Owners	Nonowners
Subadult Wing Length (± SD)	120.7 ± 3.0 (31)	119.7* ± 2.4 (172)
Adult Wing Length (± SD)	125.2 ± 3.0 (71)	124.9 ± 3.1 (29)
	124.4 ± 2.3 (71)	124.9 ± 3.0 (25)

Comparison of: (Reference: Shutler & Weatherhead 1991a)

	Territory Owners	Shallow Floaters	Deep Floaters
Wing Length (± SD)	122.1 ± 3.4 (56)	121.4 ± 3.4 (23)	120.9 ± 2.8 (35)
Tarsus Length (± SD)	25.6 ± 1.3 (56)	25.4 ± 1.1 (23)	25.3 ± 1.3 (35)
Mass (± SD)	68.7 ± 3.4 (56)	68.6 ± 2.9 (23)	68.5 ± 4.1 (35)

NOTE: Each study tests the relationship between body size and mating success in male red-winged blackbirds. Sample sizes are in parentheses.
*P < 0.05.
**P < 0.01.

and large harem size bring about higher testosterone levels, rather than vice versa (Beletsky et al. 1992).

CONCLUSIONS

None of the sexually dimorphic traits in red-winged blackbirds seem to be subject to strong sexual selection in progress, at least not according to the measures of sexual selection that have so far been used. The best case in favor of some current sexual selection can be made with respect to wing length, the one trait that has been examined most extensively. Here the overall impression is that larger males have some advantage in mating success but that the effects are quite small and inconsistent. Fewer studies are available on the other dimorphic traits, but there is nothing to contradict the conclusion that current sexual selection is weak, at best. We should emphasize that this conclusion applies only to those components of sexual selection that have been measured so far, which have to do with success in intrasexual competition for territories. The situation is still open with respect to intersexual selection in the context of copulation.

Subadults versus Adults

As we have just seen, attempts to explain why some adult males succeed and some fail in intrasexual competition have not been very successful. There is, however, one group of males that consistently experience poor success and whose lack of success does seem explicable, namely subadults. Subadult male redwings do seem capable of breeding. Wright and Wright (1944) found that testis size in subadults increases markedly in the early spring, just as it does in adults, though the increase occurs approximately two weeks later in the subadults. At the maximum, testis size in subadults was only about two-thirds that of adults, but all subadults showed active spermatogenesis, indicating that they were capable of fertilizing females. Subadult males have been found holding territories on which females nested and laid fertile clutches (e.g., Searcy 1979b), though as far as we know there are as yet no genetic data proving that subadult males sire young.

Although yearlings appear capable of mating, they experience much lower mating success than do adults. This applies to the few subadults that obtain a territory as well as to subadults in general. In a three-year study, Searcy (1979b) observed yearling males holding territories in only one year. In that year, the six yearlings with territories had a mean harem size of 2.3, compared to a mean of 5.9 for 43 adults; the difference was significant (P < 0.01). The poor success of these subadults might be explained by their having held inferior territories, by their being themselves less attractive to females, or some combination of both. Existing evidence indicates that females normally choose breeding situations primarily on the basis of territory quality. The evidence comes, however, from situations in which females choose among territories all owned by adult males, and thus does not bear on the question of whether females discriminate against yearling males. Whatever the basis of female choice, a number of observed differences between subadults and adults were found by Searcy (1979b) that could explain the difference in harem sizes: (1) mean wing lengths were significantly smaller ($P < 0.01$) for subadults (125.0 mm) than for adults (131.3 mm); (2) subadults had significantly smaller, more orange epaulets; and (3) subadults, naturally, were younger and thus less experienced. The mean mass of the subadults, however, was no lower than that of the adults, and the subadults were just as aggressive as were adults toward a caged male placed on their territories.

If subadults with territories have poor mating success, it is much more typical for subadults to fail to obtain a territory altogether and thus experience zero success. Certainly a very high percentage of subadults lack a territory, in many populations approaching 100%. The rarity of territory ownership by subadults compared to adults can in part be explained by the fact that all subadults lack the advantage of prior ownership of a territory. A more illuminating way to examine subadult success is to compare the success of subadults and adults in obtaining a territory for the first time. We can make this comparison using our male cohort reconstructed from Orians and Beletsky's (1989) data on ages at which males first obtained territories (Box 7.1). According to this reconstruction, if we consider those males who have not held territory previously, 36.2% of the two-year-olds succeed in obtaining a territory, 42.5% of the three-year-olds, and 64.0% of the four-year-olds, compared to only 2.5% of the subadults. Subadults thus have a much lower probability of obtaining a territory than do older males that are similarly seeking their first territories.

The failure of subadults to obtain a territory can again be attributed to a number of factors. Shutler and Weatherhead (1991a) found one subadult among 57 territory owners, two subadults among 25 shallow floaters, and 10 subadults among 45 deep floaters. The 13 subadults in this sample were compared to 114 adults males and proved to have significantly shorter wing lengths, smaller epaulets, and lower masses (Table 7.6). The subadults did not differ from adults in tarsus length, condition, or bill dimensions. Five subadults were tested for dominance in aviaries, and these tended to lose most of their encounters with adults, but the trend was of doubtful significance because of the small sample size (Shutler and Weatherhead 1991a). Other aviary experiments have provided clearer evidence that subadults are dominated by adults in aviaries (Wiley and Harnett 1976, Searcy 1979d).

We conclude that subadult male redwings have very low success in com-

Table 7.6

Comparison of the Morphology of Adult and One-year-old Males in an
Ontario Population.

Measure	One-year-olds (13)	Adults (114)	P
Length of Red (mm)	35.1 ± 0.9	37.4 ± 2.1	<0.001
Length of Yellow (mm)	39.4 ± 2.8	43.2 ± 2.6	<0.001
Wing Length (mm)	117.4 ± 3.8	121.3 ± 3.1	<0.001
Tarsus Length (mm)	25.7 ± 0.7	25.5 ± 1.4	NS
Bill Length (mm)	23.0 ± 0.6	22.9 ± 1.1	NS
Bill Depth (mm)	11.9 ± 0.6	11.6 ± 0.6	NS
Mass (grams)	64.7 ± 5.1	68.5 ± 3.5	<0.05
Condition (grams/mm^3 × 10^5)	4.1 ± 0.3	3.8 ± 0.3	NS

SOURCE: Shutler and Weatherhead (1991a).
NOTE: Values are means ± SD. Sample sizes are in parentheses. Probability levels are based on
t-tests comparing values corrected for seasonal changes in morphology.

petition for territory and that this in turn is the major cause of their low
mating success. Within adults, we had difficulty in pointing to any trait that
differed between successful and unsuccessful males, but in comparing subadults to adults, we have the opposite problem: there are too many differences between subadults and adults, so that we cannot say which is most
responsible for the failure of subadults. These differences include smaller
size (as measured by wing length); smaller, less brilliant epaulets; less experience; and younger age. Note that these are all differences between subadult
and adult males, not between successful and unsuccessful subadults.

The only study comparing successful and unsuccessful subadult males is
that of Weatherhead et al. (1987), who found that subadults holding territories were significantly larger (in wing lengths) than subadults from the general population (Table 7.5). This opens the possibility that there is current
sexual selection acting on subadult males; however, the overall effect must
be small since such a low proportion of subadults obtain territories and
mate. More work is needed here, replicating Weatherhead et al.'s (1987)
comparison of size in successful and unsuccessful subadults, and investigating other traits as well.

Counterselection

Even if we thought there was strong sexual selection in progress favoring the
extreme of some trait, we would not necessarily assume that the trait was
evolving toward that extreme; instead we would be more likely to hypothesize that there was some other selective force favoring the opposite extreme
and thus keeping the trait stable. A hypothesis of strong sexual selection in

progress thus goes hand in hand with a hypothesis of strong counterselection. Usually the counterselective force would be assumed to be survival selection. This scenario of survival selection balancing sexual selection has been tested in red-winged blackbirds with respect to body size. Selander (1965, 1972) proposed that sexual selection for large size in male birds in general would be balanced by survival selection favoring smaller size. Large size might have a cost in decreased survival either because large males would be farther from the energetic optimum and thus more subject to starvation and other types of condition-dependent mortality, or because large males would be more vulnerable to predation.

Selander's hypothesis predicts that current survival selection should favor small males. Seven data sets are available to test this prediction with respect to male red-winged blackbirds (Table 7.7).

RETURNING VERSUS NONRETURNING MALES

In each of three years, Searcy (1979c) compared the wing lengths of males returning from their wintering grounds to his study site to those of males not returning. Some of the males that did not return may have survived and dispersed off the study site, but this source of error should be quite small due to the strong site fidelity of adult male redwings. Returning males were smaller in two years and larger in the third; the differences were all small and none were significant (Table 7.7).

LIVING VERSUS DEAD MALES

Johnson et al. (1980) compared sizes of males that were captured alive in a winter roost in Texas to sizes of males found dead. Both wing and tarsus lengths were measured. One problem with these data is that a bias was found toward measuring the same individuals as larger when dead than when alive; the authors attempted to correct for this bias by adding the difference to the measurements of live birds (2.358 mm for wing lengths, 0.709 mm for tarsus lengths). The corrected data showed that for both adults and subadults, dead males were slightly smaller than live males in both wing lengths and tarsus lengths, though the difference was statistically significant only for adult tarsus lengths (Table 7.7).

SURVIVING VERSUS KILLED MALES

Weatherhead et al. (1984) compared sizes of surviving males to sizes of males that died following spraying with a surfactant at an early spring roost in Quebec. The presumed cause of death was hypothermia. Wing lengths were not taken because it was feared that the surfactant might change feather morphology. Among the measurements that were taken, ulna lengths are probably most closely proportional to wing lengths, and these did not differ between survivors and nonsurvivors for either adult or subadult males (Table

Table 7.7
Survival Selection on Male Body Size in Red-winged Blackbirds.

Measure (mm)	Age Class	Survivors[a]		Controls[b]		P	Reference
Wing Length	Adults	127.9	(17)	128.5	(25)	NS	Searcy 1979c
		131.0	(20)	131.6	(21)	NS	
		130.8	(25)	130.2	(10)	NS	
	Adults	127.9	(24)	116.7	(279)	NS	Johnson et al. 1980
	Subadults	122.2	(60)	122.0	(154)	NS	
	Adults	124.9	(25)	124.0	(169)	NS	Weatherhead et al. 1987
	Subadults	119.7	(172)	119.9	(208)	NS	
	Adults (museum)	121.2	(61)	121.0	(14)	NS	Weatherhead & Clark 1994
	Adults (field)	124.0	(47)	122.7	(160)	0.001	
	Subadults (field)	119.3	(19)	118.4	(329)	NS	
Tarsus Length	Adults	31.0	(24)	29.2	(279)	0.005	Johnson et al. 1980
	Subadults	29.6	(60)	29.3	(154)	NS	
Ulna Length	Adults	34.4	(7)	34.4	(11)	NS	Weatherhead et al. 1984
	Subadults	34.2	(26)	34.1	(12)	NS	
PCI	Adults (museum)	0.10	(55)	−0.22	(11)	NS	Weatherhead & Clark 1994
	Adults (field)	0.73	(47)	0.59	(149)	NS	
	Subadults (field)	−0.44	(19)	−0.41	(274)	NS	

[a]Survivors are those known to have survived over a given period.
[b]Controls are either a sample of those known to have been alive at the start of the period (Weatherhead et al. 1987, Weatherhead and Clark 1994) or a sample of those known or presumed to have died during the period (Searcy 1979c, Johnson et al. 1980, Weatherhead et al. 1984). Sample sizes are in parentheses.

7.7). Survivors and nonsurvivors also did not differ in humerus length, nor in the first principal component derived from 11 morphological measurements.

AUTUMN VERSUS SPRING MALES

Weatherhead et al. (1987) compared wing lengths of individuals captured at a roost during autumn and those captured at the same roost the following spring. The roost was located in Quebec, near the northern limit of the species' distribution, so it could be argued that the inhabitants of the roost at both seasons were local residents rather than transient migrants, and thus that any change in size between the autumn and spring samples would have to be due to differential overwinter survival. Mean wing lengths did not differ significantly between the two samples for either adult or subadult males (Table 7.7). There was, however, significant stabilizing selection on young males (against both extremes) and significant disruptive selection on adults (in favor of both extremes).

SIZE AND AGE-AT-DEATH

During a six-year study of an Indiana population, Yasukawa (1987) measured wing lengths of two-year-old males and then determined their subsequent ages at death. Again, he assumed that any adult male disappearing had died rather than dispersed. The correlation between wing length and lifespan for 17 males was negative ($r = -0.438$) but marginally nonsignificant ($P < 0.10$ by a two-tailed test).

AUTUMN VERSUS SPRING MUSEUM SPECIMENS

Weatherhead and Clark (1994) compared sizes of museum specimens that had been captured in Ontario in autumn to sizes of those that had been captured in spring. Adult males that had survived to spring were virtually identical in wing lengths to the autumn (control) males and were slightly but nonsignificantly larger in a principal component measure of size (Table 7.7).

SURVIVING VERSUS CONTROL MALES

Weatherhead and Clark (1994) also compared sizes of males captured in the field and seen again the subsequent year (survivors) to sizes of males captured and not seen again (controls). Adult survivors were significantly larger in wing lengths and nonsignificantly larger in a principal component measure of size than were adult controls. Subadult survivors did not differ significantly from controls in wing length and were significantly smaller than controls in a principal-component measure of size (Table 7.7).

In summary, there is no clear pattern of survival selection against large males. Only two comparisons are statistically significant, and both of these show a large male survival advantage. Ignoring statistical significance, more of the trends are toward a survival advantage of large males than toward an advantage of small males. Weatherhead et al. (1984) have argued that only data examining annual (rather than short-term) survival are appropriate to test Selander's hypothesis, which would restrict us to just Searcy's (1979c), Yasukawa's (1987), and Weatherhead and Clark's (1994) data, but this subset of studies also do not give convincing evidence of a small male advantage.

There is another way to rescue Selander's hypothesis, which is to propose that a small male survival advantage occurs during the period of rapid growth early in life, and so before the time looked at by any of the studies in Table 7.7. Size dimorphism is negligible at the time of hatching in red-winged blackbirds, but males grow more rapidly than females in the nest and are about-one third heavier at the time of fledging (Fiala 1981). Fiala and Congdon (1983) measured energy budgets of nestlings using doubly labeled water and estimated that males assimilate 1014 kilojoules during the nestling period, compared to 797 kilojoules for females. As males require more energy, they may be more susceptible to starvation than are females. Teather (1992) has shown, however, that in red-winged blackbirds male nestlings receive more food from their parents than do females, in large part because the bigger males are able to reach higher when begging for food. Blank and Nolan (1983) found that male nestlings were more likely to starve than were females, but Fiala (1981) found equal numbers of males and females starving. Sex ratios in the nestling stage are slightly but significantly biased toward fewer males (47.2% males; Weatherhead and Teather 1991), but as sex ratios do not seem to change with nestling age (Fiala 1981), the bias appears to be due to events prior to hatching, and thus before there is any differential growth in the sexes. There is no direct evidence on survival of large versus small male nestlings or fledglings in red-winged blackbirds, let alone on "genetically big" versus "genetically small" males (nestlings might be small because they are undernourished, and they might die for the same reason). Thus, survival selection against large males during the period of growth is an intriguing idea, but there is little evidence supporting the hypothesis at this time.

We conclude that there is no evidence of survival selection against large size in male red-winged blackbirds. This fits with our conclusion that there is not strong sexual selection in progress favoring large size; if there is little sexual selection in favor of large size, we would not expect strong counter-selection in favor of small size either.

Why Current Sexual Selection Is Weak

We have seen that there is a great deal of variation in mating success among male red-winged blackbirds, and thus a strong opportunity for sexual selection. Nevertheless, the bulk of the evidence suggests that current sexual selection acting on male redwings is weak, meaning that the opportunity for sexual selection is not translated into actual sexual selection. This conclusion must be tempered because of two gaps in our knowledge: one concerning female choice in the context of copulation, and one concerning intrasexual selection acting on subadult males. Still, we do know that there are large differences in mating success between those adult males that succeed in obtaining a territory and those that fail, and also substantial differences in mating success among adult territory owners according to their harem sizes, and yet these differences do not translate into sexual selection. Why not? Below we present two partial explanations, one based on resident's advantage and lottery competition and the other on self-limiting sexual selection.

RESIDENT'S ADVANTAGE AND LOTTERY COMPETITION

There are two extreme ways one can picture a territorial system such as that of male red-winged blackbirds: as a "meritocracy" or a "lottery." In the meritocracy model, only able, deserving males get territories, and the very best males of all get the really attractive territories with the largest harems. The meritocracy is maintained because of constant competition for territory, in which any inferior owner is thrown out by an abler nonowner, and exceptionally able owners leave poor territories to take better ones. Ability is determined by traits contributing to resource holding power (RHP), and because of the success of the able males there is strong current sexual selection in favor of these traits. In the contrasting lottery model, owners have an advantage simply from being owners, and cannot be displaced by nonowners even if the latter have far greater ability. The principal method of becoming an owner is to find a vacancy, where the old owner has died or left voluntarily. The first nonowner to arrive at the vacancy fills it, and nonowners take whatever vacancy they find, rather than waiting for one commensurate with their abilities. Once a male has a territory, he seldom attempts to move to a better one. Chance, then, determines which males succeed and how successful they are, and because there are only chance differences between successful and unsuccessful males, there is little or no current sexual selection.

A key distinction between the lottery and meritocracy models lies in the extent to which resident's advantage as opposed to RHP explains success in maintaining territory ownership. If only males of high RHP are able to maintain territories, then merit will be rewarded; if there is a strong resident's advantage, then whatever male gets to a vacancy first will be able to take and maintain the territory, and lottery competition will prevail. In chapter 3 we gave some evidence that territory owners are almost always able to defeat intruders on the territory, in other words, that owners show "site dominance." What we need to know is whether site dominance is due to superior RHP or to a resident's advantage.

A number of hypotheses have been proposed to explain site dominance in redwings and other species, each of which proposes some asymmetry between owners and nonowners (Beletsky and Orians 1987b). The first possibility is an RHP asymmetry (Parker 1974). This hypothesis proposes that RHP has determined who has succeeded in obtaining a territory, so that owners on average have much higher RHP than nonowners. As a consequence, owners are usually able to defeat nonowners. A second possibility is a value asymmetry, meaning that the territory has greater value to owner than nonowner. Such an asymmetry might arise because the owner's experience with the territory increases its value to him. Game theory models indicate that settling contests using a value asymmetry can be an evolutionarily stable strategy (ESS) even if the opponents are otherwise equal (Maynard Smith and Parker 1976). The third and final possibility is an arbitrary asymmetry (also called an uncorrelated asymmetry), meaning an asymmetry that is unrelated to either RHP or value. Again, game theory indicates that using an arbitrary rule, such as "owner wins," to settle contests can be an ESS under certain conditions (Maynard Smith and Parker 1976, Maynard Smith 1979). Either an arbitrary asymmetry or a value asymmetry produces a resident's advantage; residents have an advantage just because they are residents. If RHP asymmetries explain site dominance, then there is no resident's advantage. Thus, the important distinction in the present context is between RHP asymmetry on the one hand and value and arbitrary asymmetries on the other.

We have already presented considerable evidence against the RHP hypothesis, namely the lack of systematic differences between owners and nonowners in such characteristics as size, aggressiveness, and display traits. Additional evidence against the RHP hypothesis is provided by Shutler and Weatherhead (1991a). These authors housed together in aviaries different classes of males captured on the same territories: original owners, shallow floaters, and deep floaters. The outcome of aggressive interactions was used to determine dominance relationships between different pairs (dyads). Among adults, original owners were dominant over floaters in just 19 of 35 dyads (54%), not significantly more than half. Original owners won in six of 11 dyads (55%) with shallow floaters and in 13 of 24 (52%) with deep floaters. Shallow floaters were dominant over deep floaters in only 13 of 25 dyads. Thus, there was no evidence that winners in competition for territory were superior to losers in dominance.

These aviary experiments seem particularly significant. Yasukawa (1979), Eckert and Weatherhead (1987a), and Shutler and Weatherhead (1991a) all compared winners and losers in competition for territories, using a variety of methods and focusing on a variety of male traits. Little evidence was found of consistent differences between adult winners and losers on any trait, but the possibility remains that the important male traits were simply overlooked

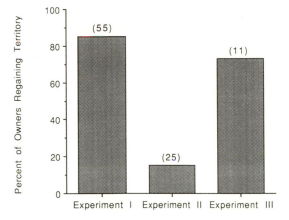

Figure 7.2 Percentages of owners regaining their territories after being held in captivity, based on data in Beletsky and Orians (1987b, 1989a). Owners were held either 1–2 days (Experiment I) or 6–7 days (Experiments II and III). Owners on release contended with replacements that had been present either up to two days (Experiments I and III) or up to seven days (Experiment II). The crucial factor affecting the success of the owners was not their own time in captivity but rather the occupancy time of their replacements.

by the researchers. Shutler and Weatherhead's aviary experiment showed, however, that success in aggressive encounters among captives does not predict success in competition for territories; therefore, whatever traits determine RHP among captives, those traits must not determine success in competition for territory.

Other studies have attempted to test the RHP hypothesis against one or more of the alternatives. Beletsky and Orians (1987b) removed owners, held them for 7 to 49 hours before releasing them, and then observed their attempts to reclaim their territories. Overall, 47 of 55 original owners succeeded in reclaiming all or part of their territories (Figure 7.2). This could be taken as showing that the original owners had especially high RHP, enabling them to evict the replacements; however, in most cases the defeated replacements were neighboring territorial males and thus original owners themselves. Out of nine cases where previously nonterritorial males became replacements, the old owners evicted them completely in only four. This result does not support the hypothesis that owners have higher RHP than nonowners, but there is an out, namely that owners lost RHP while in captivity. In fact, males did tend to lose mass while captive (though not very much), and males unsuccessful in evicting replacements had lost more mass on average than successful males.

In a subsequent set of experiments, Beletsky and Orians (1989a) removed owners and held them for seven days. Only four of 25 of these males (16%)

succeeded in reclaiming their territories after release (Figure 7.2), significantly fewer than the 47 of 55 that regained their territories (85%) after one to two days in captivity in the previous experiments. There are two possible explanations for the diminished success of original owners after seven days in captivity: first, that they lost more RHP during seven days in captivity than did males held two days, and second, that their replacements gained an advantage by holding the territory for a longer period before being challenged. Beletsky and Orians (1989a) devised an ingenious test of these two possibilities, a "double removal" experiment in which the original owner was removed and held for seven days and the replacement male was removed two days before the original owner was released. The original owner then faced a new replacement male that had held the territory only one or two days. In this experiment, eight of 11 of the original owners (73%) regained their territories (Figure 7.2), compared to four of 25 (16%) in the corresponding single-removal experiment. This result argues that the crucial factor in determining the success of original owners in reclaiming their territories is not the length of time they have been held captive but the length of time their replacements have occupied the territory.

Which of our asymmetries might change during the replacement's time of occupation? Neither RHP nor the arbitrary asymmetry of ownership seem likely possibilities; not RHP because it should not change at all, and not ownership because it seems likely to be as fully established at two days as at seven. Instead, value asymmetry seems the likeliest possibility, as it does seem plausible that knowledge of the territory might increase appreciably between two and seven days of occupation (Beletsky and Orians 1989a).

Shutler and Weatherhead (1992) performed similar experiments, removing owners and holding them for two to 30 days, meanwhile removing up to nine replacements. The intent was to test possible sources of value asymmetries. One possibility examined was that value asymmetries stem from familiarity with neighbors, that is, that territories are more valuable to owners who are familiar with their neighbors than to nonowners who are not. Owner success in reclaiming the territory, however, was not affected by the number of neighbors the territory had. A second possibility was that territories were more valuable to owners because they had invested in offspring and/or pair bonds on the territory. Results were not consistent with this hypothesis either; owner success did not differ between males that had or had not sired young on the territory, nor did owner success vary with harem size. In fact, the basic evidence for value asymmetries was lacking in this study, in that the success of replacements in retaining the territory was not correlated with their time in occupation. This last result differs from what Beletsky and Orians (1989a) found. Shutler and Weatherhead (1992) concluded that the best explanation for site dominance in their population is the arbitrary asymmetry hypothesis.

Although the two most relevant studies differ on whether they support the value asymmetry hypothesis (Beletsky and Orians 1987b, 1989a) or the arbitrary asymmetry hypothesis (Shutler and Weatherhead 1992), the important thing from the viewpoint of the present chapter is that these and other studies agree in rejecting the RHP hypothesis. Thus, what separates owners from nonowners is not, for the most part, intrinsic traits that affect RHP, but rather what we might call the accident of ownership.

Another crucial distinction between the meritocracy and lottery models is in the methods by which territories are usually obtained: either by aggressive takeover from an existing owner (meritocracy) or by occupation of a vacancy (lottery). We reviewed evidence on initial occupations earlier (chapter 3). In one study, 78% of new owners obtained their territories by filling a vacancy (Yasukawa 1979), whereas in a second study, 85% of new owners obtained their territories this way (Picman 1987). These observations again support the lottery model.

A final distinction between the meritocracy and lottery models concerns the frequency with which territory owners "trade up," leaving their current territory to take a better one. This distinction is not as clear as the previous two, in that trading up would not be necessary in a meritocracy model if the initial territories obtained by males matched their abilities. Conversely, considerable trading up is compatible with a lottery model, as long as those males trading up do so by filling vacancies rather than by deposing other males. Male red-winged blackbirds in fact do show strong fidelity to their territories, remaining on the same territory in successive years in about 90% of cases (chapter 3). Beletsky and Orians (1987a) found, however, that those males who do move tend to be ones who experienced low reproductive success in the year before moving. In addition, males tended to move to new territories that were superior to their old territories, as judged by previous production of young on the territory, although this trend was not significant. Trading up does occur, then, but at a low rate, and it is not known whether it is usually accomplished by aggressive takeovers or filling vacancies.

Although at the beginning of our careers we assumed the redwing system was a meritocracy, we have gradually come to believe it more closely resembles a lottery. Others have expressed similar views (Eckert and Weatherhead 1987c). Support for the lottery model includes the lack of consistent differences between winners and losers in competition for territory, the observation that most males obtain territories by filling vacancies, and especially the evidence supporting a resident's advantage. As we have argued, lottery competition weakens current sexual selection, in that success is determined by chance rather than by male traits. We should be careful, however, not to overemphasize the random, lottery elements of the system; there is room for some deterministic effects as well. Even though most males obtain territories that are vacant, there are nonetheless cases of active takeovers, where a new owner defeats and throws out the old. When vacancies open up, there may be more competition among floaters to fill the vacancy than we are aware of, in which RHP might play a key role. These remaining openings for deterministic effects would certainly repay greater study, starting simply with detailed observations of the events that occur when vacant territories are filled.

SELF-LIMITING SEXUAL SELECTION

As we stated before, biologists usually picture sexual selection being balanced by survival selection: sexual selection favors one extreme in a trait, and the trait is stabilized by survival selection favoring the other extreme. An alternative to this picture is self-limiting sexual selection. Here sexual selection is originally directional, causing the trait to evolve toward an extreme. As the trait changes, sexual selection changes also, coming to favor an intermediate in the trait's new range of values. At this point, sexual selection itself acts to stabilize the trait.

A hypothesis of self-limiting sexual selection was applied by Searcy (1979c) to male body size in red-winged blackbirds. The hypothesis assumes that there is a size for males that is optimum for energy balance, but because large size is advantageous in fights over territory, male size has been driven above the energetic optimum by sexual selection. Once mean size is above the optimum, larger males are farther from the optimum than are smaller males, and so are at a disadvantage in terms of energy balance. There are a number of reasons to think that energy balance is important in competition for territory. First, many takeovers take place via wars of attrition, in which males must sustain chasing and display over prolonged periods. Second, males often must defend their territories for weeks or months before females even begin to settle, at times of the season when temperatures are still quite low and food is relatively hard to find. Third, throughout the period of defense, owners must perform displays and chases, all of which consume energy. Small males may have an advantage in these energy-costly aspects of territory defense which would tend to balance their disadvantage in fights. The result would be that sexual selection no longer favors larger and larger male size but rather acts to stabilize size.

This hypothesis makes a number of predictions that can be tested. One prediction is that males should be energy-limited during territory defense. As evidence of this, Searcy (1979c) showed that territorial males in a Washington population lost mass during the breeding season, something also found in other studies (Shutler and Weatherhead 1991a). In addition, Searcy (1979c) showed that males provided with extra food on their territories spent more time on territory, and had higher display rates while there, than did control males. The hypothesis also predicts that large males are more en-ergy-limited than small males. As one test, Searcy (1979c) used data on geographic variation to predict the mass of males given their wing lengths and then showed that, during the breeding season, males that were large in terms of wing length were farther below their expected mass than were small males. In addition, he showed that display rates tended to be lower in larger males. The self-limiting sexual selection hypothesis also fits with the finding that there is no counterselection acting on size through survival selection.

Although the evidence above supports the hypothesis of self-limiting sex-ual selection, there is one problem: there is no evidence that sexual selection favors males of intermediate size. Instead, sexual selection seems to us to favor large size, though only weakly. We therefore propose a modified ver-sion of the hypothesis, as illustrated in Figure 7.3. Here, mating success increases with size over some range, but the relationship eventually asymp-totes. The asymptote occurs because males of larger and larger size are increasingly energy limited, and energy balance is important in competition for territory and mates. Adult sizes are currently at or near the asymptote, where larger-than-average males have only a minor mating advantage. As

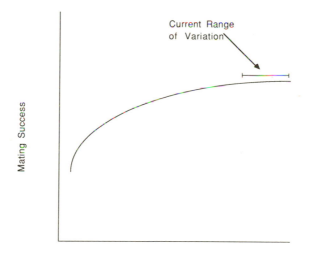

Body Size

Figure 7.3 A modified version of the hypothesis of self-limiting sexual selection on size in red-winged blackbirds. Mating success increases with size because of the advantage of large size in fighting for territories, but the relationship asymptotes because males of larger and larger size are increasingly energy limited, and energy balance is also impor-tant in competition for territories and mates. Adult sizes are currently at or near the asymptote, so larger-than-average males have little advantage in competition for territo-ries and mates.

sexual selection is at least somewhat stabilizing in this scenario, strong survival selection for small size is not necessary, which fits with the data on size and survival.

Conclusions

There is clearly a great deal of variation in reproductive success among male red-winged blackbirds, much of which is attributable to variation in mating success. This produces a strong opportunity for sexual selection. Available evidence indicates that this opportunity for sexual selection is for the most part not translated into current sexual selection. Tests for sexual selection in progress have concentrated on sexually dimorphic traits, such as visual and vocal display, size, and aggressiveness. These tests have used two main approaches, one in which males that are successful in obtaining a territory are compared to those that are not, and a second in which traits of territory owners are correlated with their harem sizes. Both approaches give results that are for the most part negative. The best case for any sexual selection in progress can be made for wing length; here 15 tests have been made, 11 of which showed trends toward a large male advantage, only three of which were statistically significant. Thus, even for size, the evidence is consistent with, at best, quite weak sexual selection in progress.

If we accept that harem sizes are determined largely by territory quality, then the tests that have been made for sexual selection in progress have focused primarily on intrasexual selection. Intersexual selection, in terms of female choice at copulation, may or may not be acting more powerfully. We have suggested two reasons why intrasexual selection is currently weak in red-winged blackbirds. One is that competition for territory follows a lottery model: males obtain territories by being lucky enough to find a vacancy (i.e., by being in the right place at the right time), and once in place they are difficult to dislodge due to resident's advantage. Because chance plays a large role in success, systematic differences between successful and unsuccessful males are minimized. The second reason for weak intrasexual selection is that sexual selection may be to some degree self-limiting. Specifically, sexual selection for large size in males may be self-limiting because the advantages of large size in fights may be balanced by a disadvantage in energy-demanding aspects of competition for territories.

8 Adaptations for Sexual Selection

An adaptation is "a characteristic of an organism whose form is the result of selection in a particular functional context" (West-Eberhard 1992). An adaptation for sexual selection, then, is a characteristic whose form is the result of sexual selection. In this chapter we consider evidence that certain traits of male red-winged blackbirds are in this sense adaptations for sexual selection.

A variety of methods can be used to test whether a trait is an adaptation (Curio 1973, West-Eberhard 1992). One is to compare "naturally occurring variants" as to their "efficiency or reproductive success" (West-Eberhard 1992). This is essentially the method used in testing for selection in progress. A demonstration of sexual selection in progress acting on a trait is thus one type of evidence that the trait is a sexually-selected adaptation. This evidence is not necessary, however, in that a trait can be a sexually-selected adaptation without being subject to current sexual selection (Grafen 1988). As we saw in the last chapter, there is little evidence for sexual selection in progress acting on male traits in red-winged blackbirds.

A second method of testing for adaptation is to correlate the form of a trait in various species with differences in the environments or other circumstances of those species (West-Eberhard 1992). This is the "comparative" method. In comparative tests of sexual selection, the usual procedure is to correlate the form of a trait in various species with the mating system of those species. The test assumes that the strength of sexual selection varies with mating system, in particular being greater under polygyny than monogamy. The logic of the test is that if the trait is a sexually-selected adaptation, it ought to be more exaggerated in polygynous than monogamous species. One problem with this logic is that although variance in mating success may be greater in polygynous systems, it does not follow that sexual selection is necessarily stronger (Clutton-Brock 1988). The comparative method also suffers from the usual problems of correlational methods in inferring causation; in particular, any association found between a trait and mating system may be due to an association of both with a third variable, such as phylogeny.

A third method of testing for adaptation is to investigate the pattern of use of the trait. The logic here is that if a trait is used primarily in contexts in which mating success is determined, this is evidence that the trait is an adaptation for sexual selection. For example, if a visual trait is displayed only to conspecific females during courtship, or only to conspecific males during aggressive encounters, a sexual-selection function would be supported; conversely, if the trait was displayed only to predators, this would suggest a predator-deterrence function. As Andersson (1994) has remarked, "quantification of the context in which a trait is used is a powerful means of revealing its function."

A fourth and final method of testing for adaptation is to alter the trait experimentally "in order to see how this affects its efficiency in a particular function or environmental condition" (West-Eberhard 1992). This is the method we prefer, as it is the only experimental rather than correlational method of studying adaptation. Others have agreed with us in giving primacy to this type of evidence (Grafen 1988). This is not to say that a watertight case for adaptation can be made using this method alone. One might alter a trait experimentally, and demonstrate in this way that the trait has a certain effect on (for example) mating success, and yet this effect might not be the function in the sense of being the reason the trait evolved (Williams 1966). Overall, the best demonstration of adaptation can be made by combining evidence from experimental manipulation with some of the other types of evidence listed above.

In this chapter we review evidence on whether male traits in red-winged blackbirds are sexually-selected adaptations, using results from the comparative method, from patterns of use, and from experimental manipulations. The four categories of male traits examined are, again, the major categories of sexually dimorphic traits: visual display, vocal display, size, and aggressiveness. We will consider evidence for sexually-selected adaptation not only in the major traits within each category, but also in more subtle traits such as delayed plumage maturation, song repertoires, and song-recognition mechanisms.

Visual Display

EPAULETS: PATTERNS OF USE

Many of the display postures of male red-winged blackbirds serve to expose and exaggerate the males' red-and-yellow epaulets (see chapter 3). The most common such display is the song spread. Observational evidence indicates that the song spread is an important component of territory defense. Song spreads occur spontaneously as part of advertisement behavior, in response to potential intruders flying over the territory, and in response to trespassers perched within the territory (Nero 1956b, Orians and Christman 1968). Song spreads are given at especially high rates during portions of the breeding season when trespassing is most common (Peek 1971). Males display their epaulets to females as well as to males, and in fact the highest-intensity song spreads are more likely to be directed at females (Peek 1972). Purely court-ship displays (crouch and precopulatory display) also feature the epaulets prominently.

Yasukawa (1978) presented a taxidermic mount together with recorded song to territorial male red-winged blackbirds and observed whether or not the subjects attacked the mount within 15 minutes. Attackers differed from nonattackers in a number of aspects of song-spread display: attackers gave song spreads of higher mean intensities, gave medium-intensity displays at higher rates, and gave low-intensity displays at lower rates than nonat-tackers. Thus, the degree and frequency with which males expose their epaulets in song spread serve to signal aggressive intentions toward other males.

Exposure of the epaulets is not limited to specific display movements such as the song spread, however. Males also expose and even flare their epaulets when not singing and while otherwise maintaining a normal resting posture (Nero 1956b). Such exposure of epaulets by a resting male is usually done only by territory owners while on their own territories.

The contexts in which epaulets are displayed suggest that they are used as intersexual signals in courtship and intrasexual signals in territory defense. Epaulets are not normally displayed in other contexts, such as foraging or predator avoidance. We conclude that patterns of use support the hypothesis that epaulets are sexually-selected adaptations.

Epaulets: Experimental Manipulations

Experimental evidence on the function of epaulets in territory defense comes from studies in which the epaulets were blackened in selected territorial males and effects on competition for territory were observed. Two such studies were published almost simultaneously by Peek (1972) and Smith (1972), followed later by a third study by Morris (1975). The results of these three independent experiments agree in showing that epaulets play an important role in territory defense.

Peek (1972), working in a small marsh in Pennsylvania, blackened epau-lets with magic markers or dye and also produced some white-winged males by clipping the tips of the red and yellow feathers to expose their white bases. Of 17 males whose epaulets were obliterated, six (35%) subsequently lost their territories, compared to none of 13 control males (Figure 8.1). Another six of the experimental males were classified as maintaining their territories only with difficulty, compared to none of the controls. All 12 of

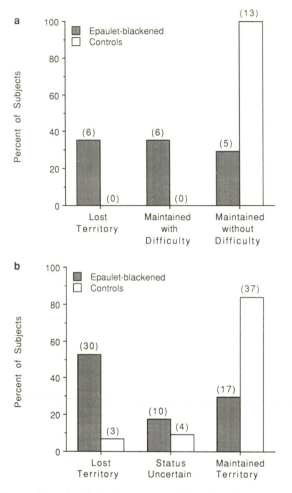

Figure 8.1 Effects of epaulet-blackening on the ability of territory owners to maintain their territories. Results in (a) from Peek (1972), in (b) from Smith (1972). In both studies, males with blackened epaulets were much more likely to lose control of their territories than were control males. Numbers of males in each category are shown above the bars.

the experimental males that lost their territories or maintained them with difficulty were ones whose epaulets were treated in the premating period; by contrast, the five males whose epaulets were obliterated in the nesting period all maintained territories without difficulty. Peek (1972) observed much higher mean rates of trespassing by conspecific males on territories of black-winged males (5.8 trespasses/hour, N = 4) and white-winged males (7.7 trespasses/hour, N = 2) than on territories of controls (0.9 trespasses/hour,

N = 7). Trespassing males tended to remain on the territories of experimental males for long periods, ignoring the displays of the owners. Peek concluded that epaulets are important in territory defense, especially for close-in interactions with other males.

Smith's (1972) experiments were performed in marshes in Massachusetts and Washington. Experimental males had their epaulets blackened with dye or paint. Control males were captured and treated in a similar fashion, except that their epaulets were merely wiped with alcohol, which had no effect on coloration. Experimental and control males were held for similar periods, in all cases less than 25 minutes. Of 57 experimental males, 30 (53%) lost their territories and 17 (30%) maintained them, whereas the status of the remaining 10 was uncertain (Figure 8.1). In contrast, of 44 control males, only three (7%) lost their territories, whereas 37 (84%) maintained them and four had uncertain status (Figure 8.1). Unlike Peek (1972), Smith found no statistically significant relationship between the timing of epaulet-blackening and the likelihood of territory loss. He did have the impression, however, that black-winged males defending marginal territories were more successful at maintaining ownership than black-winged males defending territories of higher quality, perhaps because competition for marginal territories was less intense.

Morris (1975) conducted a small study in Ohio in which she blackened the epaulets of six males, using an additional six as controls. Two of the experimental males lost their territories, as did two of the controls. Trespass rates were somewhat higher on territories of experimental males than on those of the controls.

Taken together, the epaulet-blackening experiments demonstrate that the epaulet is an important component of territory defense in red-winged blackbirds. Although black-winged males were able to sing and otherwise display, their lack of epaulets seemed to reduce considerably their ability to prevent trespassing by other male redwings. In addition, trespassing males often did not respond to the displays and chases of the black-winged males, so that trespassers remained on territories for long periods. Perhaps because of their inability to discourage trespassing, black-winged males were much more likely than controls to lose their territories. As we have seen, territorial males have much higher mating success than do nonterritorial males, so loss of territory carries a severe sexual-selection cost.

Although the epaulet-blackening experiments indicate an intrasexual function of epaulets, they are less clear on an intersexual function. In Smith's (1972) experiment, most males whose epaulets were blackened lost their territories, but some were able to retain at least some area, and these males were able to attract nesting females. In contrast, Peek (1972) found that epaulet-blackened males who successfully retained their territories were unable to attract mates. The latter result may indicate that females discriminate during settlement against males without epaulets, but another explanation is that females discriminate against males who are experiencing difficulty in defending their territories.

More direct evidence of an intersexual function of epaulets comes from an unpublished study by David A. Enstrom (pers. comm.). Enstrom gave captive female red-winged blackbirds a choice of associating with control males possessing normal epaulets or with experimental males possessing altered epaulets. In two sets of hour-long trials, 16 untreated and 18 estradiol-implanted females spent significantly more time near control males than near males with epaulets reduced in size by half through blackening (Trial 1: 16.1 versus 11.3 minutes; Trial 2: 30.3 versus 17.3 minutes; $P < 0.05$ for each trial according to Wilcoxon signed-ranks tests). In a third set of trials, 11 estradiol-implanted females spent more time near males with artificially brightened epaulets than near control males, but the difference was not statistically significant (27.3 versus 22.8 minutes; $P > 0.10$ according to a Wilcoxon signed-ranks test). Whether preferences for association reflect mating preferences remains to be demonstrated, but for now we regard these results as supporting an intersexual effect of epaulets.

Although the field experiments on epaulet blackening make it clear that epaulets are favored by intrasexual selection, they leave some doubt about the exact mechanism. One possibility is that epaulets are essential to species recognition, so that black-winged males lose their territories because they are not recognized by rival males as conspecifics. The alternative is that variation in epaulet size and color serves as some sort of intraspecific social signal, for example of dominance status ("status signaling"), aggressive intentions, or territory ownership. Under this social-signaling hypothesis, black-winged males are recognized as conspecifics but are not viewed by intruders as serious competitors for the territory, and so are ignored.

One obvious experiment to test between the species-recognition and social-signaling hypotheses is to enlarge the epaulets of territorial males, as enlargement beyond the normal range cannot enhance species recognition but should enhance the social signal. Unfortunately, altering epaulet size on live males turns out to be difficult. Røskaft and Rohwer (1987), however, have carried out an ingenious alternative experiment that gives us at least some of the essential evidence. These authors removed territory owners and set out taxidermic mounts on their now-vacant territories. The taxidermic mounts had either blackened epaulets, normal epaulets, or epaulets tripled in size by gluing on epaulets taken from other redwing skins. All mounts were stuffed in a song-spread posture and were fastened to a T-shaped bar. An empty bar served as a control. In one experiment, six territory owners were removed from a long, narrow marsh, and the three mounts and the control bar were placed, equally spaced, on the center portion of the marsh, each "defending" a strip 40 meters long (see Figure 8.2). Røskaft and Rohwer

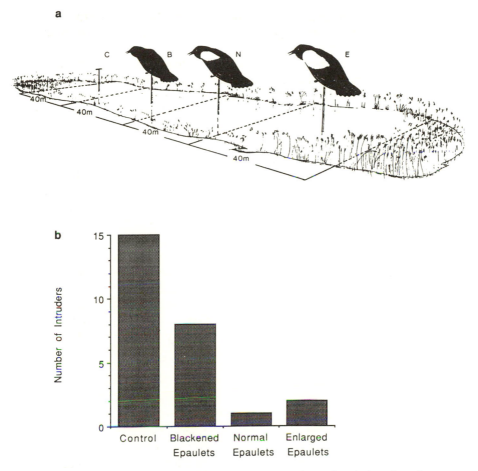

Figure 8.2 Effects of epaulet size of defenders on numbers of male intruders, from Røskaft and Rohwer (1987). (a) The original defenders were removed from the marsh, and mounts were placed out in the above order; later mount positions were reversed. The "enlarged" mount had epaulets tripled in size; the "normal" mount had natural epaulets; the "blackened" mount had blackened epaulets; and the control treatment was an empty perch. (b) The number of intruders that landed in each section of marsh. Significantly more intruders landed in sections with control and blackened treatments than in sections with enlarged and normal treatments (P < 0.001).

then observed on which strips intruders landed over the course of two evenings and one morning. Mounts were switched between strips occasionally to control for any area effects. As floaters moved into the marsh and started to establish their own territories, they in turn were removed. In all, 23 male redwings first landed nearest the black-winged mount or the control, whereas only three landed nearest the normal or enlarged-epaulet mount.

Thus, prospecting males showed a significant tendency to avoid mounts with normal or enlarged epaulets.

A second experiment tested whether intruders discriminate between normal and enlarged-epaulet mounts. This experiment was done on a series of small marshes, each originally defended by a single male. Røskaft and Rohwer began a trial by removing the territory owner and then placing the normal mount on one half of his vacant territory and the enlarged-epaulet male on the other. Subsequently, intruding males showed a significant tendency to avoid the enlarged-epaulet mount, with only seven males landing first on the area defended by that mount, compared to 20 that landed first on the area defended by the normal mount. The eight intruders who remained on the experimental marshes and began to show territorial behavior continued to avoid the enlarged-epaulet mount, spending only a mean of 1.9 minutes closer to that mount compared to 14.3 minutes closer to the normal mount; this difference was also statistically significant. All eight of these males directed their first bill-up displays at the normal mount rather than at the enlarged-epaulet mount.

It seems clear, then, that increasing the size of a mount's epaulets enhances its effectiveness in territory defense. This result does not seem compatible with the species-recognition hypothesis: tripling the size of the epaulets, thus putting them well outside the species' normal range of variation, cannot be argued to enhance species recognition. The result does, however, seem fully compatible with the social-signaling hypothesis: if variation in epaulet size and color signals varying levels of threat or dominance, then supernormal epaulets ought to convey a particularly intense signal and thus be especially intimidating to competitors.

Further evidence against the species-recognition hypothesis comes from experiments by Hansen and Rohwer (1986) testing the ability of male redwings to recognize epaulet-blackened males as conspecifics. Here, territory-owning males were confronted with either taxidermic mounts of conspecific males with their epaulets blackened or with mounts of male Brewer's blackbirds. Male Brewer's blackbirds were chosen because they resemble epaulet-blackened males more closely than do any other birds in the Washington study area. Nevertheless, the territory owners showed clear evidence of species recognition, attacking the epaulet-blackened redwings far more often than the Brewer's blackbirds. This result argues that epaulets are not necessary for species recognition.

If we accept the social-signaling hypothesis, we are left with the question of what exactly epaulet size and color signal. Again, one possibility is social status. This possibility is supported by correlations between dominance status and epaulet size found in captive flocks. Eckert and Weatherhead (1987d) formed three captive groups of adult males, with 28 subjects altogether. In pairwise comparisons, males with larger epaulets dominated those with smaller epaulets in 71.7% of all dyads, significantly more often than chance. A second possibility is that exposure of the epaulets signals aggressiveness; this possibility is supported by the correlations Yasukawa (1978) found between song-spread intensity and attack likelihood in territorial males. A third possibility is that exposure of the epaulets signals territory ownership; this possibility is supported by evidence showing that intruders entering occupied territories normally cover their epaulets, whereas males entering territories whose owners have been removed very soon start exposing their epaulets, as they begin to assert ownership (Hansen and Rohwer 1986). We believe that these three types of signaling are all mutually compatible and that most likely all three are operating at once. We conclude that the epaulets of male red-winged blackbirds function in male-male competition for territories and that this function comes about because epaulets serve as a signal of high dominance status, territory ownership, and aggressiveness.

COVERABLE BADGES

Signals of status that are relatively cheap to produce and maintain have been termed "badges" (Dawkins and Krebs 1978, Rowher 1982). "Coverable badges" are simply badges that can be readily concealed (Hansen and Rohwer 1986). The epaulets of male red-winged blackbirds are badges in that they signal social status (territory ownership and/or dominance), and they are clearly coverable in that they can be (and frequently are) covered over and hidden by the black over-wing coverts. We have seen (above) how possession of the badge is of benefit in intrasexual competition; here we address the question of why it is advantageous to be able to cover the badge.

Hansen and Rohwer (1986) have proposed one general answer. These authors suggest that in social systems like that of red-winged blackbirds, males often trespass onto a territory for reasons other than to challenge the owner. During such intrusions, it is advantageous for the intruder to cover his badge, so as to signal to the owner that he is not a threat and will leave without fighting. Thus, the benefit of being able to cover the badge is that it allows a male to enter the territories of other males without being attacked. As a test of this hypothesis, Hansen and Rohwer (1986) measured the aggressiveness of territory owners toward taxidermic mounts with: (1) completely blackened epaulets, (2) half-blackened epaulets, (3) normal epaulets, or (4) epaulets doubled in size. As predicted, the amount of aggression shown by the owners increased with the size of the epaulet exhibited by the mount (Figure 8.3).

Metz and Weatherhead (1992) tested the effects of uncovering the coverable badge by clipping the over-wing coverts on 20 territory owners, thus preventing these males from covering their epaulets. Only three of the 20 clipped males lost their territories, compared to five of 20 control males; clipping, then, did not increase the likelihood of losing ownership. In addi-

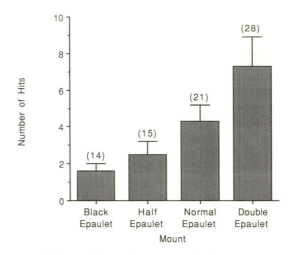

Figure 8.3 Number of hits made by territory owners (in 10 minutes) on mounts of male red-winged blackbirds with epaulets of differing sizes, based on data in Hansen and Rohwer (1986). Epaulets were doubled in size ("double") by taking epaulets from other redwing skins and gluing them on the mounts behind the original epaulets. Epaulets were halved in size ("half") or obliterated completely ("black") using black hair dye. Aggression of owners toward a mount increased with the size of the mount's epaulets. The variation in hits across epaulet-size treatment is significant according to a Kruskall-Wallis one-way ANOVA (P < 0.02). Error bars show standard errors; sample sizes are in parentheses.

tion, clipped and control males incurred nearly identical rates of trespassing (4.3 and 4.5 trespasses/hour, respectively). The one sign that clipped males had any difficulty in defense was that when neighboring males intruded, chases by clipped males were of significantly longer duration (49.5 seconds) than those by control males (19.8 seconds). Clipping did seem to have a major effect on the ability of males to trespass on other territories: clipped males were observed trespassing only three times, whereas control males were observed on other territories 16 times.

We conclude that in the context of male-male competition, the principal advantage of being able to cover the badge is in allowing males to trespass on territories owned by other males with less likelihood of being attacked, as Hansen and Rohwer (1986) hypothesized. An advantage to territory owners when on their own territories seems more doubtful. It may well be, however, that covering the badge also benefits a male by reducing his chances of attracting the attention of a predator, both on and off his territory.

COLOR BANDS

A great many studies of redwing behavior, including our own, have used colored "leg" bands to enable researchers to recognize individual birds without having to recapture them (actually, the bands are placed on what is the bird's foot anatomically). For many years, researchers used this technique without considering that color bands might affect the behavior of the banded birds and of the birds with which they interact. This all changed when studies by Nancy Burley and colleagues with zebra finches showed that band color affected mate choice by both males and females (e.g., Burley et al. 1982). It became important, therefore, to find out whether color banding was distorting the results of field studies of red-winged blackbirds. A second reason for studying color bands is to use them as a way of manipulating visual signals; for example, giving a male a number of red bands can be interpreted as enhancing whatever signal his red epaulet makes, thus providing a method of investigating what that signal is.

The most direct way of addressing the methodological issue is to analyze the effects on male success of the band combinations actually used to facilitate individual recognition. Two such analyses have been published, both of which have concentrated on red bands, as red seems the color most likely to interfere with the redwing's own signaling system and is a band color known to have effects in other species (Burley et al. 1982, Hagan and Reed 1988). Beletsky and Orians (1989c) performed a retrospective analysis of 11 years of data from a Washington population in which males were given three to five color bands, made of either plastic or anodized aluminum. These authors found that males given one or more red bands did not differ from those given no red bands in harem size, fledging success, or years of breeding (Table 8.1). Weatherhead et al. (1991) similarly analyzed four years of data on an Ontario population in which males were given three aluminum color bands. Red bands had no significant effects on harem sizes, fledging success, or return rates of males (Table 8.1). Data from this and another population indicated that red bands also had no effect on the likelihood that a male would lose his territory (Weatherhead et al. 1991).

Table 8.1
Effects of Red Color Bands on Success of Male Red-winged Blackbirds in Holding Territory, Attracting Females, and Breeding.

Measure	Males with Red Band(s)	Males with No Red Bands	P	Reference
Harem Size	4.1 ± 2.4 (560)	4.1 ± 2.4 (231)	NS	Beletsky & Orians 1989c
Number of Fledglings	5.2 ± 6.1 (560)	5.3 ± 6.0 (231)	NS	
Years Breeding	2.2 ± 1.4 (193)	2.0 ± 1.2 (76)	NS	
Harem Size	2.3 ± 0.9 (75)	2.0 ± 0.9 (44)	NS	Weatherhead et al. 1991
Number of Fledglings	3.6 ± 3.0 (68)	2.5 ± 2.4 (41)	NS	
Return Rate	52.9% (68)	68.6% (35)	NS	

NOTE: Means ± SD are given. Sample sizes are in parentheses.

In a study designed to test the signaling effects of color bands, Metz and Weatherhead (1991) augmented the amount of red displayed by certain male redwings by attaching five red plastic leg bands to each male. Control males were given five black bands, which match the natural leg color. Eleven of the 19 red-banded males subsequently lost their territories, compared to none of 19 black-banded males. The 11 red-banded males that lost their territories lost them to four black-banded and seven unbanded males. Loss of territory seemed to be associated with increased harassment by neighbors. Of 93 cases in which red-banded males chased an intruder off the territory, the intruder was a neighbor in 36 (39%); by comparison, neighbors were involved in only 22 of 89 chases (25%) by black-banded males ($P < 0.05$).

One interpretation of Metz and Weatherhead's (1991) results is that because red is perceived as a signal of aggression and dominance by other male redwings, red-banded males were regarded as signaling higher status than they were entitled to, and were consequently harassed to the extent that many had to abandon defense (Metz and Weatherhead 1991). A contrasting interpretation is that red-banded males were regarded as abnormal (Metz and Weatherhead 1991), perhaps as being heterospecific, and so were ignored by competitors. In an unpublished study, Metz and Weatherhead (pers. comm.) attempted to test these interpretations by presenting to territory owners models of male red-winged blackbirds bearing six red, six black, or six blue bands. Owners were less aggressive toward red-banded models than toward black-banded or blue-banded ones. Given the fact that owners are more aggressive toward a model the larger is that model's epaulets (Hansen and Rohwer 1986), we think this result weighs against the interpretation of red bands as representing an exaggerated aggressive signal; certainly, red bands are not treated in the same way as are red epaulets.

Metz and Weatherhead (1991) found that males given five red color bands had significantly smaller harem sizes (mean = 1.2 females) than males given black bands (mean = 1.8 females). One explanation of this result is that females discriminate against males with many red bands because such males seem abnormal. A second explanation is that females discriminate against red-banded males solely because such males are experiencing difficulties maintaining their territories.

The results of Metz and Weatherhead's (1991) experiment, in which males given red bands were more likely to lose territories, seem to conflict with those of Beletsky and Orians (1989c) and Weatherhead et al. (1991), in which there was no effect of red bands on territory tenure. The explanation may lie in the fact that the red-banded males in Metz and Weatherhead's (1991) experiment were given substantially more red bands than were the red-banded males in the other two studies. Each red band may have a small effect that accumulates as more red bands are added, or there may be some threshold number of red bands necessary to see any effect. Because of the possibility that red bands have small, cumulative effects, it would seem prudent to minimize use of red bands in future studies (Weatherhead et al. 1991), despite the negative results of the retrospective analyses of effects of red bands on male success and the ease with which such bands can be seen by ornithologists.

DELAYED PLUMAGE MATURATION

Another interesting aspect of visual display in male red-winged blackbirds is that males in their first potential breeding season, at the age of one year, have a different plumage than adult males. Yearling males are generally black, but with variable amounts of brown. Their epaulets are more orange than red and are often flecked with brown or black. The plumage of yearling males ("subadult plumage") is thus intermediate between the plumage of females and the definitive male ("adult") plumage. In addition, yearling males are much more variable in plumage than are adult males. As we have stressed before, yearling males are reproductively competent (Wright and Wright 1944, Payne 1969) and sometimes do hold territories and acquire mates (chapter 7). Such a pattern, in which reproductively competent but young males exhibit a predefinitive plumage, is common among passerines (Rohwer 1978) and has been named "delayed plumage maturation."

Hypotheses to explain the evolution of delayed plumage maturation can be divided into two classes: those proposing winter advantages and those proposing summer breeding-season advantages. Subadult plumage may be advantageous in the winter because it signals low status and thereby reduces intraspecific aggression, or because it is cryptic and so reduces predation (Rohwer and Butcher 1988). If winter-advantage hypotheses are to explain summer plumage, they must also assume a "molt constraint," i.e., that a spring molt is costly enough so that the winter plumage is retained into the breeding season even though it is maladaptive in that season (Rohwer and Butcher 1988). If the yearling plumage of male red-winged blackbirds is constrained in this way, then sexual selection has had little influence on the evolution and maintenance of delayed plumage maturation in redwings.

Hypotheses that explain delayed plumage maturation in terms of summer advantages usually assume that subadult plumage puts young males at some disadvantage in competition for resources and mates, and propose some compensating advantage to balance this cost. Possible compensating advantages include reduction of predation through crypticity (Selander 1965, Rohwer et al. 1980), reduction of intraspecific aggression via signaling of low status (Lyon and Montgomerie 1986), mimicry of female plumage (Rohwer 1978, Rohwer et al. 1980), or mimicry of juvenal plumage (Foster 1987).

We know of no data bearing on the molt-constraint hypothesis for red-

winged blackbirds. Nor do we know of any data testing the effects of yearling plumage on the probability of predation, whether during the breeding
season or outside it. There are, however, data testing whether yearling
plumage reduces intraspecific aggression. These data come from experiments by Rohwer (1978) in which he tested the aggressiveness shown by
territorial male red-winged blackbirds to mounts of adult male red-winged
blackbirds, adult females, yearling males with plumage resembling adult
males ("studly" yearlings), and yearling males with plumage closer to that of
adult females ("unstudly" yearlings). Rohwer found that territorial males
were most aggressive toward adult males and least aggressive toward adult
females and showed intermediate aggression toward yearling males. Further,
territorial males were more aggressive toward studly yearlings than toward
unstudly ones. For example, the subjects spent a mean of 41.2% of the trial
periods on (and usually pecking) the adult male mounts, 23.2% on the studly
yearlings, 12.4% on the unstudly yearlings, and 0% on the female mount.

Rohwer's (1978) results support the hypothesized summer advantage of
yearling plumage in reducing aggression from adult males. It is not clear
whether the effect occurs because yearling males are interpreted as being
females, juvenals, or simply subordinate males; as the alternatives depend
on knowing what other redwings "think" when they see yearling plumage,
they are perhaps not very testable. At any rate, the identification of one
summer advantage of delayed plumage maturation should not stop the search
for explanations of the phenomenon; other summer advantages, as well as
molt constraints, may also be operating.

Vocal Display

SONG: PATTERNS OF USE

The song of male birds has long been thought to function in the defense of
territories against other males. Howard (1920) summarized the observational
evidence for this function in his classic *Territory in Bird Life*: attainment of
full song during the breeding season often coincides with territory establishment, males do not sing when temporarily absent from their territories, and
males increase singing when an intruder enters the territory. Note that these
are all arguments based on patterns of use. Some of Howard's generaliza-
tions hold for male red-winged blackbirds, and some do not. Male redwings
sing outside the breeding season as well as during it (Orians and Christman
1968), so song is not confined to periods of territory defense. During the
breeding season, territory owners often leave the territory for short periods
and usually do not sing when absent from it. Male redwings sometimes,
though not always, increase their singing in response to an intruding male;
for example, we found that males in New York showed a significant in-

crease in singing rates in response to a simulated intrusion, but males in Pennsylvania did not (Searcy and Yasukawa 1990).

The second major function of male song, in birds in general, is thought to be the attraction of females. Howard (1920) gave as the principal evidence for this function the observation that in many species male singing rates drop dramatically as soon as the males pair. Later authors demonstrated that males in some species will resume vigorous singing if their mates are removed (Catchpole 1973, Wasserman 1977). These observations are certainly in accord with the mate-attraction hypothesis, but there are alternative explanations; for example, it may be that song functions only in territory defense and that males sing less upon pairing because, once paired, they switch their attention from territory defense to mate guarding and parental duties.

For red-winged blackbirds, it is not clear that the mate-attraction hypothesis predicts a decrease in singing after pairing, because there is such a strong possibility that additional females could be attracted after the first one has settled. In fact, there is a tendency for male singing rates to decline as the breeding season progresses (Shutler and Weatherhead 1991b), though the decline seems to occur gradually and not as a step function as in many monogamous species. Rather surprisingly, male redwings do respond to the removal of a female in the same way as do males in monogamous species. When Searcy (1988a) removed the first-settling female from randomly chosen territories, the male redwings owning those territories subsequently sang nearly twice as much as males on control territories (Figure 8.4). This is in

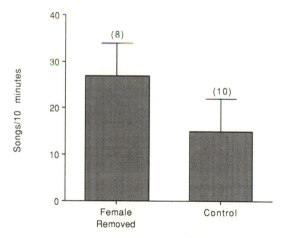

Figure 8.4 Effect of the removal of the primary female on the song rate of the male territory owner, based on data in Searcy (1988a). Shown are mean song rates (songs/10 minutes) plus a standard deviation. Primary females were removed soon after they had settled on the territories, and male song rates were monitored during the 11 days following the removals. Male song rates were significantly higher on removal territories (P < 0.01 by a Mann-Whitney U-test), which is indirect evidence that males use song to attract new females.

accord with an intersexual function of song. Also in accord with an intersexual function is the observation that males sing frequently when actively courting a receptive female.

SONG: EXPERIMENTAL EVIDENCE

The first experimental evidence on the function of song in red-winged blackbirds came from Peek (1972), who captured male redwings on their territories in the spring and removed sections of the hypoglossal nerves, which innervate the syrinx. The experimental males were subsequently unable to vocalize normally (although they could utter wheezing noises), i.e., they were "muted." Controls were also captured, anesthetized, and cut open, but their hypoglossal nerves were only touched rather than sectioned. Overall, 13 of 23 muted males (57%) suffered complete loss of their territories, compared to none of eight control males (Figure 8.5, Experiment I). The effects of muting varied with time of season. Twelve of 13 males muted in the prenesting period lost their territories, whereas only one of 10 males muted in the nesting period lost his territory. Six of the controls were treated during the prenesting period and two during the nesting period. The results indicate that the ability to vocalize is very important to the defense of a territory, at least early in the breeding season.

Smith (1976) performed a similar experiment, muting territory owners by sectioning their hypoglossal nerves. Smith's results were quite different from Peek's: only one of 15 muted males (7%) lost his territory, compared to three of 15 sham-operated controls (20%) (Figure 8.5, Experiment II). Smith's study was carried out in the same kind of habitat as Peek's (small, freshwater marshes) and covered about the same periods of the breeding season, so these factors seem unlikely to explain the difference in results. One change in methods that might have affected the outcome is that Smith used new anesthetics that gave shorter recovery times. Peek reported that both muted and sham-operated males in his experiment remained absent from their territories for several hours after treatment, long enough for replacement males to have taken over in many cases. Thus, it may be that muted males are able to retain territories when they have uninterrupted ownership (Smith's experiment), but they cannot displace a replacement owner that has taken over in their absence (Peek's experiment). This interpretation still suggests that vocalizations are important in competition for territory; however, Smith (1976) cautions that males whose hypoglossal nerves have been sectioned may experience respiratory problems, which provides an alternative explanation for their frequent loss of territory in Peek's study.

Smith (1979) introduced a new method of muting males, designed to prevent any respiratory complications. This method involves puncturing the interclavicular air sac, which sits just below the skin between the "arms" of

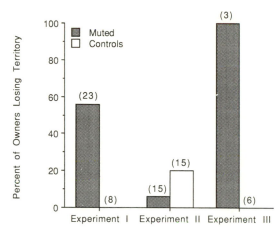

Figure 8.5 Effects of muting males on their ability to defend territories. Experiment I: results of Peek's (1972) experiments in which he sectioned the hypoglossal nerves, leading to the syrinx. Males muted in this fashion were more likely to lose territory than were controls. Experiment II: results of Smith's (1976) experiments in which males were again muted by sectioning the hypoglossals. Here muting did not increase the likelihood of losing territory. Experiment III: results of Smith's (1979) experiments in which males were muted by puncture of their interclavicular air sac. Muted males were more likely to lose territory than were controls; in this study, loss of territory was partial rather than complete.

the wishbone. The puncture releases the pressure that normally pushes the sound-producing tympaniform membranes into the air flow of the trachea. Male redwings treated in this way were rendered mute (i.e., unable to sing the species-typical song) for two to three weeks, after which they regained the ability to produce normal songs. An advantage of this procedure is that anesthesia is not necessary, and males are therefore able to reoccupy their territories immediately after release. Smith devocalized three males using this method, and all three lost substantial portions of their territory during the period when they were mute (Figure 8.5, Experiment III). When they regained their ability to sing, each of the three regained at least part of the lost territory. One sham-operated male enlarged his territory after treatment, whereas five untreated males experienced little change in their territory boundaries (Figure 8.5).

In this last muting experiment, territory size decreases and increases in parallel with the ability to sing, providing particularly good evidence that song contributes to territory defense (though the sample size admittedly is small). A remaining difficulty is that muting deprives males of the ability to produce all vocalizations, including calls as well as song, so there is a possibility that the effects of muting are due to failure to produce one of these other vocalizations. In this respect, muting experiments are nicely comple-

Table 8.2
Effects of Song Playback on Male Intrusion onto Territories during Speaker-Occupation Experiments.

Playback Type	Measure	Playback	Control	P[a]
Single Song Type (16)	Fly-through Rate	9.8 ± 6.1	11.9 ± 6.2	0.05
	Nonneighbor Trespass Rate	1.4 ± 1.4	1.4 ± 1.2	NS
Repertoire (16)	Fly-through Rate	11.4 ± 12.8	14.5 ± 14.5	0.01
	Nonneighbor Trespass Rate	1.1 ± 1.4	2.6 ± 2.4	0.01

SOURCE: Yasukawa (1981c).

NOTE: Territory owners were removed and speakers set out in their places. Observations were made during one-hour periods of silence (controls), playback of single song types, and playback of a repertoire of eight song types. Mean rates per hour ± SD are given.

[a]P values are from Wilcoxon signed ranks tests.

mented by a second experimental design, the "speaker occupation experiment" (Krebs et al. 1978). Here a male is removed from his territory, and a speaker is set out in his place, from which songs can be broadcast. If there is less intrusion onto the territory when song is played than when it is not, the conclusion that song has an effect on territory defense seems inescapable.

Yasukawa (1981b, c) performed a speaker-occupation experiment with red-winged blackbirds. After removing a territory owner, he placed two speakers on the territory and staged hour-long playback and silent control periods, in random order. During playback periods, a tape containing either a single song type or eight song types was broadcast, with the songs switching between the two speakers every five minutes to simulate an owner moving about on his territory. Yasukawa kept track of two types of intrusions onto the "speaker territory": trespasses (in which the intruder perched on the territory) and fly throughs. Playback of single song types caused a significant decrease in fly-through rates relative to control periods but had no effect on rates of trespassing by nonneighbors (Table 8.2). Playback of eight types caused significant decreases in both measures of intrusion (Table 8.2). These results reinforce the conclusion that song functions in territory defense.

One way to test the intersexual function of song with experimental data is to examine the mating success of muted males. The trouble with this test is that even if muted males prove to have reduced mating success, this could be ascribed to the male-male effects of song; that is, muted males might attract few females only because they were experiencing difficulty maintaining their territories. In Peek's (1972) experiment, three muted males held territory through part of the period in which females were settling. These three males were unmated when muted and continued unmated through the season. Unfortunately, the significance of this result is unclear because Peek does not give mating success for any control group. In Smith's (1979) second muting experiment, it is unclear how successful males were in acquiring new females while muted, and again no control group was provided.

The most satisfactory data are from Smith's (1976) first muting experiment. Here, five of the males that were muted in 1972 returned in 1973 and held territory on the study site. The 1973 "songs" of these males were more normal than their 1972 postoperative utterances, but the 1973 "songs" still lay well outside the normal range of song variation in red-winged blackbirds (Smith 1976). Despite producing abnormal songs throughout 1973, the five males attracted as many females (mean = 1.4 females/male) as did six males with normal songs (mean = 1.3).

Laboratory tests of female reaction to song playback provide a second test for an intersexual function of song. Searcy (1988b) treated captive female red-winged blackbirds with estradiol implants and then played them bouts of male redwing songs. Songs of male song sparrows were used as a control. Females responded to some of the songs with copulation solicitation display, a display given by females in nature before and during copulation. Responses were scored on a 0-to-3 scale according to the intensity and duration of the display given for each song and then were summed over the bout of songs. Display scores averaged much higher for playback of redwing songs than for control songs (Figure 8.6).

The laboratory results on solicitation display demonstrate that male song definitely does have effects on female mating behavior, stimulating females to court and copulate. Thus, these results are strong evidence of an intersex-

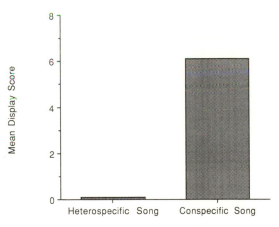

Figure 8.6 Courtship response of female red-winged blackbirds to playback of either conspecific or heterospecific song, from Searcy (1988b). Fifteen estradiol-implanted, captive females were tested twice each for response to 32 song bouts of red-winged blackbird and song sparrow song. Courtship scores measure the number and intensities of copulation solicitation displays. Females responded significantly more strongly to the redwing song (P < 0.01 by a Wilcoxon Matched-Pairs test). From *Acta XIX Congressus Internationalis Ornithologici*, vol. 1. Published for National Museum of Natural Sciences (Canadian Museum of Nature) by University of Ottawa Press, 1988. Reproduced courtesy Canadian Museum of Nature, Ottawa, Canada.

ual function of song. Given the ability of some "muted" males with radically altered songs to attract females (Smith 1976), it seems doubtful that song has much effect on female settlement, though further investigation of this point would be warranted. Instead it seems more likely that song affects the components of male mating success having to do with success in intra- and extrapair copulation, as the laboratory playback experiments simulate the events of courtship and copulation fairly well.

SPECIES RECOGNITION

Both sexes of red-winged blackbirds are able to discriminate own-species from other-species songs (Brenowitz 1981, Searcy 1988a). The mechanisms by which such discrimination is accomplished have themselves been shaped by sexual selection. Mechanisms of species recognition have been investigated in male red-winged blackbirds using the classic technique of territorial playback (Falls 1992), in which songs are played to breeding males on their territories, and one or more aspects of their aggressive response are measured. For females, the question has been addressed using the solicitation-display assay discussed above. For both sexes, subjects are presented with songs in which certain acoustic cues have been altered, to test whether those cues are necessary to evoke response.

Searcy and Brenowitz (1988) argued on a priori grounds that we should not expect the results of such experiments to be identical for the two sexes. These authors pointed out that there are two types of errors that can be made in species recognition, termed Type I and Type II errors in analogy with statistical errors. A Type I error occurs when an individual rejects (i.e., fails to respond to) a stimulus that is actually from the correct species. A Type II error occurs when an individual accepts (i.e., does respond to) a stimulus that is actually from another species. In the context of courtship and copulation, in which female response is measured, females ought to be under strong sexual selection to minimize Type II errors, as these could lead to wasted, hybrid matings. Type I errors, however, would not be so costly, as females will have many opportunities to mate. Male response is measured in experiments that simulate the intrusion of another male onto the territory. In this context, a Type I error is the more costly because allowing a conspecific male to sing on the territory may lead to its loss. Type II errors would be less dangerous because these would entail only the cost of approaching and displaying to an inappropriate target. Therefore it is predicted that females will be more discriminating, in the sense that they will be more likely to reject songs as conspecific than are males.

The prediction of greater discrimination on the part of females has been tested experimentally. Brenowitz (1982a) found that male redwings respond just as aggressively to playback of a mockingbird imitation of redwing song as to playback of a true redwing song, whereas Searcy and Brenowitz (1988) found that female redwings show very strong discrimination against the imitation in courtship. Similarly, male redwings respond just as strongly to the trill portion of the song alone as to a full song (Beletsky et al. 1980, Brenowitz 1982b), whereas females court more for the full song (Searcy and Brenowitz 1988, Searcy 1990). One discrimination task has been studied at which both male and females succeed, namely discrimination between geographic variants of conspecific song (Brenowitz 1983, Searcy 1990). Interestingly, both sexes are surprisingly unaffected by the order of elements within the song; if songs are cut into parts and the order of the parts shuffled, both males and females respond just as strongly to the shuffled songs as to the originals (Beletsky et al. 1980, Searcy 1990). Thus, there is one discrimination task on which both sexes have been shown to fail (involving order of song elements), one task on which both sexes succeed (involving geographic variants), and two on which females succeed and males fail (imitation versus true song, trill versus full song). As there are no examples of tasks on which males succeed and females fail to set against the two examples where females succeed and males fail, the hypothesis of greater discrimination on the part of females is supported.

Song Repertoires

Another aspect of vocal display that may be a sexually-selected adaptation is the song repertoire. A number of hypotheses have been proposed to explain the evolution of song repertoires in birds in general (Krebs and Kroodsma 1980, Capp 1992). One suggestion is that different song types convey different messages and that males sing multiple song types so that they can communicate each of the messages. Evidence for this hypothesis in other species is that specific song types are given in specific contexts (e.g., Morse 1966, 1970) and that listeners respond differently to the playback of different song types (e.g., Ficken and Ficken 1970). Smith and Reid (1979) stated that the use of specific song types in red-winged blackbirds was not correlated with any social or environmental contexts (though they presented no data on these points). Yasukawa (1981b) found no obvious differences in the effectiveness of eight different song types in deterring trespassing by male conspecifics, but he did not perform statistical tests. Searcy (1988b) found that female redwings responded with courtship to each of four song types presented to them and that there were no significant quantitative differences in the strengths of courtship response between the four song types. It is impossible to disprove the separate-messages hypothesis altogether, but there is no evidence supporting it for red-winged blackbirds.

A second hypothesis is that repertoires function in matched countersinging, whereby a territory owner matches the song type used by another

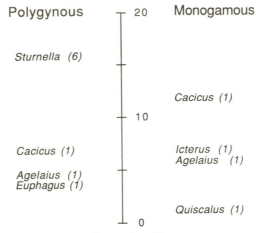

Figure 8.7 Mean song repertoire sizes for polygynous and monogamous taxa within the Icterinae, bised on data in Irwin (1990). Data are shown genus by genus. A genus is shown twice in cases where Irwin (1990) gives repertoire sizes for both a polygynous and a monogamous species within the genus. For one genus (*Sturnella*), repertoire sizes are averaged across several species (6) with the same mating system. Assignment of mating systems follows Irwin (1990). No association between mating system and repertoire size is apparent.

singer. Matching could be a signal that the singer is directing his attention to the specific male being matched (Armstrong 1973). In red-winged blackbirds, males sometimes share similar (though not identical) song types. Smith and Reid (1979) claimed that neighboring males match each other to some extent during countersinging, but again they presented no data. One way to see whether male redwings would match if they shared song types exactly is to play to a male songs from his own repertoire and note what he sings in response. Yasukawa (unpublished) conducted such a "self" song playback and found no evidence of matching above chance levels. In addition, no one has yet shown that matching is advantageous to males, in this or any other species.

This brings us to the hypothesis that repertoires have evolved due to sexual selection (Krebs and Kroodsma 1980, Catchpole 1980). Irwin (1990) provided a comparative test of this hypothesis for icterines. Her data on repertoire sizes and mating systems are illustrated in Figure 8.7. We display these data genus by genus, giving a mean repertoire size across species for the one genus (*Sturnella*) in which repertoire size has been measured in more than one species with a particular mating system. The sample size (of genera) is small, but it is clear there is not a strong trend for polygynous species to have larger song repertoires than monogamous species (Irwin

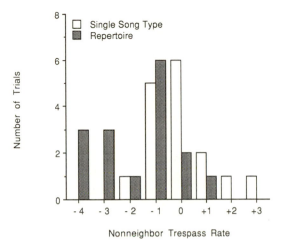

Figure 8.8 Comparison of the effects of single-song-type playback and repertoire play-back on nonneighbor trespass rates in speaker-occupation experiments, from Yasu-kawa (1981b). Histograms show the increase or decrease in trespass rates (number/hour) during playback relative to silent control periods. Decreases in trespass rates were significantly greater during repertoire playback than during single-type playbacks (P < 0.01 by a Mann-Whitney U-test).

1990). Furthermore, in the two genera with both monogamous and polygy-nous species (*Cacicus* and *Agelaius*), the monogamous species has a larger repertoire in each case. It seems unlikely, then, that variation in repertoire size between species can be explained by variation in the opportunity for sexual selection.

The hypothesis that redwing repertoires have a sexually-selected function can also be tested experimentally. A test for an effect of repertoires in intra-sexual competition is provided by Yasukawa's (1981b) speaker-occupation experiments. Here the effectiveness of playback in limiting intrusion was measured by the difference in intrusion rates between playback and silent control periods. Using this measure, playback of eight song types was not more effective than playback of single types in limiting fly throughs, but playback of the repertoire was significantly more effective in limiting tres-passes (Figure 8.8). This latter result argues that repertoires are advan-tageous in male-male competition for territories.

An experimental test for an intersexual function of repertoires was pro-vided by Searcy (1988b), who compared the reaction of estradiol-treated females to playback of single song types and repertoires. The repertoire playback consisted of 32 songs of four song types (A, B, C, and D) pre-sented in eventual variety (i.e., eight renditions of A, followed by eight Bs, then eight Cs and eight Ds). Four separate single-song-type playbacks were used, one with 32 renditions of A, one with 32 Bs, one with 32 Cs, and one

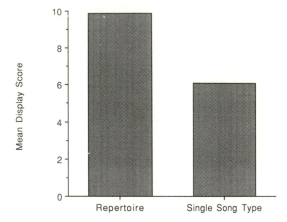

Figure 8.9 Courtship response of female red-winged blackbirds to playback of either a single song type or a repertoire of four song types, from Searcy (1988b). Eleven estradiol-implanted, captive females were tested four times each for response to 32 song bouts of either four song types or a single song type. Courtship scores measure the number and intensity of copulation solicitation displays. Females responded significantly more strongly to the repertoire of four song types (P < 0.01 by a Wilcoxon Matched-Pairs test). From *Acta XIX Congressus Internationalis Ornithologici*, vol. 1. Published for National Museum of Natural Sciences (Canadian Museum of Nature) by University of Ottawa Press, 1988. Reproduced courtesy Canadian Museum of Nature, Ottawa, Canada.

with 32 Ds. On each of four days the subjects were played the four-type tape and one of the single-type tapes, a new one of the latter each day. This design ensures that each subject hears each song type equally often in the single-type tests as in the four-type, so that any difference in response cannot be ascribed to preferences for particular song types. Display scores were significantly higher for the repertoire than for the single-type playbacks (Figure 8.9). This result argues that repertoires are more effective than single song types in stimulating females to court and copulate.

As the repertoire has an effect on both male and female listeners, it is interesting to ask which of these effects is more important as a function of the repertoire. We attempted to address this question indirectly by comparing the use of the repertoire in male-male versus male-female contexts (Searcy and Yasukawa 1990). Two separate sets of experiments were performed, one in New York and one in Pennsylvania. In each, we presented either a male stimulus (a caged male or a taxidermic mount in a song-spread posture) or a female stimulus (a taxidermic mount in a soliciting posture) on a territory and recorded the owners' vocal response. In both locales, we found a significant increase in the rate of switching between song types in response to the female stimulus (Figure 8.10). Rather to our surprise, switching rates actually decreased in response to the male stimuli; the de-

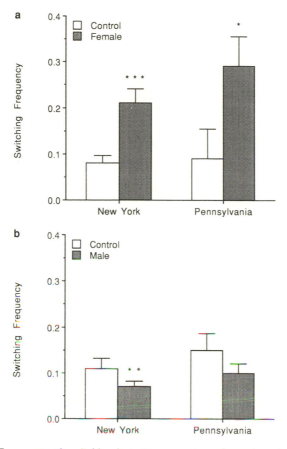

Figure 8.10 Frequency of switching between song types by male red-winged black-birds in intersexual and intrasexual contexts, from Searcy and Yasukawa (1990). Context was manipulated by presenting a live bird or a mount on a subject's territory; the stimulus was either female (a) or male (b). Switching frequency measures the proportion of opportunities to switch song types in which switches actually occurred. Half the experiments were done in New York, half in Pennsylvania. Means ± SE are shown. * indicates a significant difference at the 0.05 level, ** at the 0.01 level, and *** at the 0.001 level. Males increase switching frequency when courting females and decrease switching frequency when confronting male intruders.

crease was significant in the New York but not the Pennsylvania experiments (Figure 8.10). Thus, males seem to display their repertoire primarily to females rather than other males, suggesting that the male-female function may be primary.

The next question we consider is why repertoires have effects on females and males. Phrased another way, why have listeners (either male or female) evolved to respond to repertoires? Such response patterns may be adaptations evolved due to either natural or sexual selection. Alternatively, response behaviors may be nonadaptive.

THE EVOLUTION OF MALE RESPONSE TO REPERTOIRES

Krebs (1977) suggested that male birds might use the number of song types heard in an area as an index of the density of territorial males, and then avoid areas that were densely settled. If so, then by singing a repertoire of song types, a territory owner could exaggerate the apparent local density and thus discourage intrusion on his territory. Krebs termed this the "Beau Geste" hypothesis in reference to the novel of that name, in which French Foreign Legionnaires prop their dead comrades at the embrasures of a fort in order to deceive the enemy about the number of defenders. This mechanism does not fall neatly into an adaptive/nonadaptive dichotomy: the response of listeners in avoiding areas with many song types is assumed to be adaptive to the extent that song-type density accurately reflects true density of territorial males, but it is nonadaptive (or maladaptive) for listeners to the extent that they are deceived about density by song repertoires.

Beyond the fact that repertoires discourage trespassing, certain other aspects of redwing behavior are also consistent with the Beau Geste hypothesis. For one, singing males tend to switch song types at the same time that they change perches (Yasukawa 1981b), which should contribute to the impression that several singers occupy the territory. For another, there are strong positive correlations between the density of territorial males and the density of song types produced in an area (Yasukawa and Searcy 1985), which means that the proposed method of density assessment is workable. We have shown, however, that contrary to an assumption of the hypothesis, intruders do not avoid densely settled areas (Yasukawa and Searcy 1985). Correlations between trespass rates per territory and male density are not consistently negative, whereas correlations between trespass rates per area and male density are consistently positive. Intruders, then, do not avoid densely settled areas, which refutes a key assumption of the Beau Geste hypothesis.

Another possible explanation for the effect of song repertoires on intruders is that repertoires serve as status signals, that is, as signals of dominance status (Rohwer 1975). This hypothesis is plausible for red-winged blackbirds because repertoire size is correlated with male age and experience (Yasukawa et al. 1980), and older males do tend to dominate younger ones (Searcy 1979d). It is not known, however, whether or not intruders avoid territories owned by older males. In an unpublished study, Yasukawa and Rebecca A. Boley tape-recorded males for two consecutive hours each at Yasukawa's Wisconsin prairie and then measured the responses of the recorded males to playback of unfamiliar songs. Yasukawa and Boley found no evidence that repertoire size is correlated with aggressive response to song playback.

The remaining possibility is that the response of potential intruders to repertoires is nonadaptive. A strong argument can be made that female response to repertoires is nonadaptive (see below), so this possibility should be taken seriously with respect to males as well.

THE EVOLUTION OF FEMALE PREFERENCE FOR REPERTOIRES

Many hypotheses have been proposed to explain the evolution of female preferences for repertoires in species possessing repertoires, and a plausible argument can be made for most of these with respect to red-winged blackbirds (Searcy 1992). Some of the hypotheses suggest a natural-selective advantage for the preference. First, the preference may benefit females by helping them obtain males with better territories. This hypothesis is plausible for redwings because repertoire size is positively correlated with male experience (Yasukawa et al. 1980), and more experienced males tend to have larger and better territories (Yasukawa 1981a). Second, the preference may benefit females by helping them obtain males who will provide superior parental care. This is plausible for redwings because repertoire size is a cue to male experience, and experienced males are more likely to provision nestlings (chapter 2). Note that these first two hypotheses assume that repertoire size influences female settlement as well as willingness to copulate, which may not be the case. Third, the preference may benefit females by helping them obtain "good genes" for their offspring. Again, this hypothesis is plausible for redwings because of the correlation between repertoire size and years of breeding experience, as it can be argued that older males have demonstrated they have superior genes for viability.

A fourth hypothesis suggests the preference coevolved with the male trait in a Fisherian runaway process (Fisher 1930). Under this hypothesis, the preference evolves because of a sexual-selective advantage: sons of females with the preference possess the preferred trait and therefore experience enhanced mating success. A variety of population genetics models indicate that the coevolutionary process may work (O'Donald 1980, Lande 1981, Kirkpatrick 1982). The models assume that both the preferred trait and the preference are heritable, which has not been tested for redwing repertoires.

The above hypotheses all assume that the female preference has either evolved in response to an existing male trait and its value as a cue to territory quality, parental quality, or genetic quality, or that the female preference and male trait have coevolved. A further possibility is that the female preference existed before the male trait. For example, females might have pre-existing preferences because of properties of their sensory systems, as Ryan et al. (1990) have suggested for a frog, *Physalaemus pustulosus*. Females in this species prefer male calls with low-frequency chucks, apparently because such calls fit the tuning of the female's auditory system better than do calls with high-frequency chucks. As evidence that the female's auditory tuning predates the evolution of the chuck, Ryan et al. (1990) showed that in a second *Physalaemus* species, *coloradorum*, in which males do not produce a chuck, auditory tuning curves of females are nonetheless biased toward low-frequency calls. Thus, the female preference for low-frequency chucks may be a built-in bias of the female's sensory system, a bias that the calls of males have evolved to exploit ("sensory exploitation").

A similar explanation can be applied to redwing repertoires. First, as evidence that preferences for repertoires in females can predate the evolution of repertoires in males, Searcy (1992) showed that in another icterine, the common grackle, females respond preferentially to repertoires of song types even though males of the species sing only one song type each. The methods used were the same as in the experiments on repertoire preference in female red-winged blackbirds: captive females were implanted with estradiol and then tested for response to 32 songs of either four song types or a single song type. In each of two experiments, subjects performed more courtship display for four types than for single types. Second, it can be argued that repertoire preferences are the product of two built-in features of animal nervous systems, habituation and stimulus specificity. Habituation refers to the waning of a response during repeated stimulation. Stimulus specificity means that a response (including habituation) is specific to a particular stimulus type; if habituation is specific to a particular stimulus, then response will recover when the stimulus is changed. Both habituation and stimulus specificity are very widespread features of response systems (Hinde 1970).

Moreover, habituation and recovery can be observed in the response of female birds to repertoires. Figure 8.11 shows the mean display scores given by female redwings in response to each of the 32 songs in the repertoire preference tests (Searcy 1988b). Display scores declined throughout the single-type playbacks. Display scores also declined during repeated presentation of a single song type during a repertoire presentation, but response recovered whenever the song type was switched. Two of the three recoveries in the redwing experiment were statistically significant. Similar patterns of habituation and recovery have been found for female response to repertoires in other species, including common grackles (Searcy et al. 1982, Searcy 1992).

It is possible, then, that female preferences for repertoires simply represent built-in response biases, on which selection has not acted, and that male song repertoires have evolved to exploit these already-existing biases. Under this hypothesis, repertoires are selectively advantageous to males, whereas preferences for repertoires are selectively neutral to females. Another possibility is that the female preference originated as a response bias and was

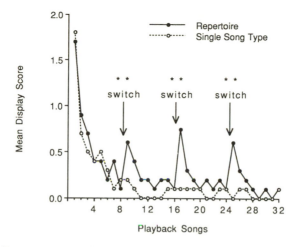

Figure 8.11 Response of female red-winged blackbirds to each song in 32 song play-backs of either a single song type or a repertoire of four song types, from Searcy (1988b). Display scores measure the intensities of the copulation solicitation postures given by the subjects. "Switch" indicates points at which the four-song-type playback switched song types. ** indicates a significant increase (P < 0.01) in response be-tween two successive songs. From *Acta XIX Congressus Internationalis Ornithologici*, vol. 1. Published for National Museum of Natural Sciences (Canadian Museum of Na-ture) by University of Ottawa Press, 1988. Reproduced courtesy Canadian Museum of Nature, Ottawa, Canada.

subsequently exaggerated in coevolution with the male trait. Under this lat-ter hypothesis, the preference originally would be selectively neutral but then would acquire a sexual-selective advantage. Testing whether any such coevolution has occurred would be extremely difficult (but see Searcy 1992).

GROWL AND PRECOPULATORY NOTE

Males produce two vocalizations besides the song that may have sexually-selected functions, the growl and the precopulatory note. The growl is a short, low-frequency, harsh vocalization. Orians and Christman (1968) noted this call occurring in three rather different situations: (1) when a male is demonstrating a nest site to a female, he may give a growl as he lands in the vegetation; (2) during aggressive interactions, the growl may be directed at another male; and (3) when a bird of another species is trespassing on the territory, the owner may direct the growl toward the trespasser. The first two contexts suggest intrasexual and intersexual functions, but no one has stud-ied the effects of the growl on listeners in any of these situations. Depending on the listener, growls may have either a courtship or an aggressive meaning (as does the song), but this is only speculation.

Precopulatory notes are single, short notes occurring somewhere in the range of 2000 to 8000 Hertz. They are repeated in series, usually at a rate of between two and 10 notes per second. Their exact form differs between males, even within a population (Searcy 1989). These notes are given by males only when courting a receptive female and are typically accompanied by the precopulatory posture. Usually, a courting male intersperses streams of precopulatory notes with occasional songs.

Searcy (1989) studied the effects of precopulatory notes on estradiol-treated, captive females. Playback of precopulatory notes was more effective in eliciting solicitation display from females than was playback of hetero-specific song, but was significantly less effective than playback of con-specific song. Mixtures of precopulatory notes and song tended to elicit more display than song alone, though the effect was statistically significant with only one of three playback tapes. Playback of precopulatory notes did not cause females to approach the speaker. It seems, then, that precopulatory notes do function to elicit sexual receptivity in females but that they are not as important as male song in this respect.

Size

Investigating size as an adaptation is hampered by certain difficulties. First, pattern-of-use data are unavailable. Males can choose to display or not to display their epaulets or songs according to context, and we can learn something about the function of these traits by analyzing the patterns. Each male is stuck with one size regardless of context, so there is no pattern to analyze. Second, experimental data are similarly unavailable. Epaulets can be blackened or enlarged, songs can be presented with altered rates or repertoire sizes, aggressiveness can be enhanced or diminished, but nothing analogous can be done with size. This leaves comparative evidence as the only means of testing the hypothesis that size is a sexually-selected adaptation.

Selander (1958) was the first to point out that monogamous species of icterines tend to be less size dimorphic than polygynous and promiscuous ones. As evidence, he provided data on percentage differences between the sexes in wing, tail, bill, and tarsus lengths in 18 species, 13 of which he was able to classify as to mating system with some confidence. Later, Selander (1972) pointed to the existence of the same trend, toward greater dimorphism in nonmonogamous species, in other groups of birds, and he gave as a general explanation that in nonmonogamous species "the number of matings a male may obtain is potentially very large, and there is an increased selective premium on characters that function in intrasexual and epigamic interactions directly related to mating." Large size, he thought, would usually be advantageous primarily in intrasexual competition, though it might sometimes be favored by female choice as well (Selander 1972).

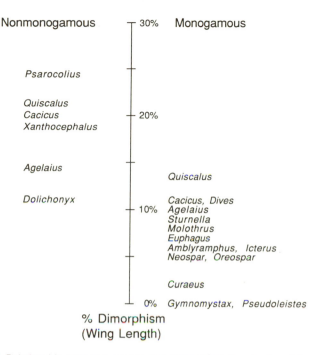

Figure 8.12 Relationship between sexual size dimorphism (in wing length) and mating system in icterines, based on data from Orians (1985). % dimorphism is the average percentage by which male wing lengths exceed female wing lengths for the species of a given genus. Three genera (*Agelaius*, *Cacicus*, and *Quiscalus*) appear twice because these genera contain both monogamous and nonmonogamous species. There is a clear trend for nonmonogamous taxa to have greater dimorphism.

Orians (1985) provided a much larger data set, with a total of 62 species of icterines classified as to mating system (monogamous or nonmonogamous) and assigned a figure for percentage dimorphism in wing length. We illustrate Orians's data in graphical form in Figure 8.12. The data are displayed genus by genus, with mean dimorphism calculated over all the species within a genus with the same mating system. A given genus occurs twice only if it contains both monogamous and nonmonogamous species, in which case we have calculated a mean dimorphism for the monogamous species and a separate mean for the nonmonogamous species. It is clear from Figure 8.12 that Orians's (1985) data do show a strong trend in the expected direction: nonmonogamous taxa are more size dimorphic, with very little overlap between the nonmonogamous and monogamous groups. Red-winged blackbirds are categorized as showing 18% dimorphism in wing length, making them more dimorphic than any of the 46 monogamous species included in Orians's (1985) data set.

One problem with the above analysis is that it does not control body size. Size dimorphism increases with body size in many families of birds as well as in other taxa (Payne 1984, Björklund 1990, Gaulin and Sailer 1984), and polygynous species often tend to be larger than related monogamous ones (Wiley 1974, Payne 1984). Therefore, the association between mating system and size dimorphism may be merely the indirect consequence of associations between mating system and size and between size and size dimorphism. A second problem is that our analysis does not control for phylogeny. The danger here is that two character states, such as polygyny and pronounced size dimorphism, may now be associated because present-day species have inherited both from ancestral taxa, rather than because their association has been brought about by some selective mechanism. Statistical methods exist for teasing out the effects of common ancestry, given a knowledge of the phylogeny of the group being examined (Ridley 1983, Bell 1989).

Webster (1992) has performed an analysis of size dimorphism and mating system in the Icterinae that controls for both the above problems. Webster measured size dimorphism as the ratio of male tarsus length to female tarsus length, and characterized mating system by mean harem sizes. A regression analysis showed quite a tight relationship between the natural log of dimorphism and the natural log of harem size for 35 species of icterines (Figure 8.13). Webster also found a positive relationship between body size (tarsus length) and the degree of dimorphism, but this relationship was considerably looser than that between dimorphism and harem size. In a multiple regression analysis, harem size was significantly associated with dimorphism, whereas body size was not. These results argue that there is a direct association between mating system and size dimorphism, independent of body size.

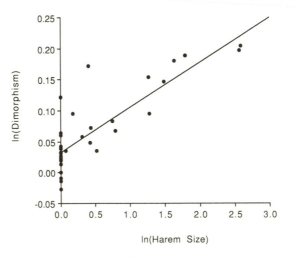

Figure 8.13 A regression of size dimorphism on harem size for 35 species of icterines, from Webster (1992). Size dimorphism is measured as the natural log of the ratio of male tarsus length to female tarsus length.

In controlling for effects of ancestry, Webster (1992) used two estimates of icterine phylogeny, one based on restriction fragment length polymorphisms (RFLP) in mitochondrial DNA (Freeman 1990) and an unpublished one constructed by Scott M. Lanyon based on cytochrome B sequences. Both harem size and dimorphism showed phylogenetic effects. Dimorphism was still positively correlated with harem size within almost all groups of related species, including those groups containing red-winged blackbirds. With either phylogeny, analysis of covariance shows that size dimorphism was still strongly associated with mating system when effects of ancestry were removed, as well as when both ancestry and body size were controlled simultaneously. Thus, size dimorphism is strongly related to harem sizes within the icterines independently of any effects of body size and phylogeny.

These results can be taken as evidence for the importance of sexual selection in causing dimorphism, assuming that mean harem sizes measure the strength of sexual selection within species. The weakness in the argument is in this last assumption. Mean harem sizes provide a measure of the opportunity for sexual selection, but opportunity for sexual selection does not necessarily translate into selection (see chapter 7). Again, the problem is that the strength of sexual selection depends not only on the variance in male mating success but also on the strength of the relationship between mating success and the trait of interest, and the heritability of that trait. Still, the existence of the interspecific trend does seem to us to be indirect evidence for the importance of sexual selection in causing size dimorphism in red-winged blackbirds as well as in other nonmonogamous icterines.

Aggressiveness

PATTERNS OF USE

Most within-species aggression performed by male red-winged blackbirds during the breeding season occurs in territory defense. Displays that seem to have an aggressive meaning, such as bill-ups, songs, and song spreads, are all at least in part territorial displays. Mildly aggressive acts between males, such as chases and displacements, occur commonly during competition for territory, though they also occur in contexts such as foraging and roosting. Knockdown, drag-out fights between males occur only during competition for territory. Pattern-of-use data thus indicate that aggression functions in competition for territory, and therefore in determining mating success.

Plasma testosterone levels increase in males as the breeding season approaches (Beletsky et al. 1989). In Washington State, the peak coincides not with the onset of territory defense but with the period of nest initiations (Beletsky et al. 1989). This is the period when trespassing, courtship, and mate guarding are maximal, so the peak in testosterone may be associated with either intersexual or intrasexual behavior.

EXPERIMENTAL MANIPULATIONS

Searcy and Wingfield (1980) attempted to manipulate levels of aggressiveness in captive male red-winged blackbirds using implants of testosterone or flutamide. Flutamide is an antiandrogren, a class of compounds that are thought to act as competitive inhibitors of androgens such as testosterone. Five groups of three males were formed, and in each group, one male was treated with testosterone, one with flutamide, and one with cholesterol (as a control). The treatments had a strong effect on aggressiveness as measured by attack likelihoods, with testosterone-treated males the most aggressive and flutamide-treated males the least. Five out of five testosterone-treated males were dominant in their groups, whereas four out of five flutamide-treated males were at the bottom of their hierarchies.

Harding et al. (1988) formed aviary groups containing intact control males, castrated males, and castrated males given one of several hormone treatments. Castration lowered levels of aggressive display such as songs and song spreads. Treatments that provided both androgenic and estrogenic metabolites were most effective in restoring these aggressive displays; testosterone was one such treatment. Testosterone-treated males were the most aggressive of any treatment group, as measured by their willingness to attack other males. Testosterone-treated males also tended to be dominant in their aviaries. The results of the two aviary studies show that androgen and antiandrogen treatments can be used to manipulate aggressiveness in red-winged blackbirds.

Two studies have used these techniques to manipulate aggressiveness in free-living male redwings. Searcy (1981) gave either testosterone, flutamide, or cholesterol implants to males already owning territories on marshes in New York. Such implants are known to continue delivering hormones for many weeks. Subjects were observed to see whether they gained or lost territory, with only large-scale gains and losses (50% or more of a territory) recorded. Combining data from two years, there was some tendency for testosterone-treated males to gain territory more often than controls (Table 8.3). There was no tendency for flutamide-treated males to do worse than control males in terms of gains and losses (Table 8.3).

Beletsky et al. (1990) gave testosterone implants to 12 floater males in their Washington population. Testosterone levels more than doubled in these males relative to other floaters, but only one of the 12 succeeded in obtaining a territory. Eleven territory owners given testosterone implants showed no tendency to expand their territories, nor did four flutamide-treated males show any tendency to lose territory. Treatment with flutamide plus ATD (1, 4, 6-androstatriene-3, 17-dione) had much greater effects, however. ATD

Table 8.3

Numbers of Males Experiencing Changes in Territory Ownership after
Treatment in the Field with Testosterone, Flutamide, or Cholesterol.

Treatment	Net Gain	No Change	Net Loss
Testosterone	5 (27.8%)	12 (66.7%)	1 (5.6%)
Cholesterol	0 (0%)	12 (80.0%)	3 (20.0%)
Flutamide	2 (25%)	3 (37.5%)	3 (37.5%)

SOURCE: Searcy (1981).

NOTE: Testosterone is thought to increase aggressiveness, flutamide is a competitive inhibitor of testosterone, and cholesterol serves as a control. Only large-scale
gains and losses, in which 50% or more of a territory changed ownership, are
included.

blocks the conversion of testosterone to estradiol, so treatment with both
flutamide and ATD blocks both the direct action of testosterone and its action through conversion to estradiol. Of seven males given both flutamide
and ATD, six lost 10 to 100% of their territories. By comparison, only one
of four control males lost part of his territory. When implants were later
removed from some of the flutamide-plus-ATD subjects, three of four who
had lost territory then regained it. Even though flutamide plus ATD lowered
success in holding territory, it was not clear that these substances affected
aggressiveness; treated males did not differ from controls in the frequency of
aggressive behaviors such as fighting, chasing, and singing.

Manipulation of hormone levels seems to have more dramatic affects on
dominance in newly formed aviary groups than on success in competition
for territories in the field. Presumably, the difference occurs because of social inertia in the field situation (Beletsky et al. 1990), meaning that relationships between territorial neighbors are well established and difficult to
change. Greater social inertia in a basically nonmigratory population may
explain why testosterone treatment had less effect in Washington (Beletsky
et al. 1990) than in New York (Searcy 1981), where the experimental manipulation was performed soon after males returned from migration. The
androgen blockers flutamide and ATD together had strong effects on success
in maintaining territories, but it is not clear that these effects occurred because of changes in aggressive behavior. Overall, we still hold to the hypothesis that male aggression functions primarily in intrasexual competition
for territory, but we must admit that the experimental evidence is not yet
conclusive.

Conclusions

We began the chapter by defining sexually-selected adaptations as traits
whose forms are the result of sexual selection. There is strong evidence that

several traits of male red-winged blackbirds are sexually-selected adaptations in this sense. The best cases can be made for visual and vocal display traits, as these traits are more easily subjected to the experimental manipulations that provide the best evidence of a trait's function. For the male's red-and-yellow epaulets, the evidence includes that (1) the epaulets are used in courtship and aggression; (2) males whose epaulets are blackened or otherwise removed have a greatly increased chance of losing their territories and thus forfeiting their opportunities to mate; and (3) mounts with enlarged epaulets perform better in keeping intruding males off a territory. The experimental data support an intrasexual function of epaulets. The pattern-of-use data indicate the epaulets also have an intersexual function, and this function is supported by experimental evidence showing that females prefer to associate with males having normal rather than reduced epaulets.

The evidence that song is a sexually-selected adaptation includes that (1) song is used in courtship and territory defense, (2) muted males are more likely to lose their territories than controls, (3) playback of songs lowers pressure by intruders on a territory, and (4) playback of song to females stimulates courtship. Song repertoires in males and species-recognition mechanisms in both sexes may also have been shaped by sexual selection.

The evidence is weaker for size, as pattern-of-use data cannot be applied to this trait, and experimental tests have not been performed. The best available test is a comparative one, and here the results are as clear as can be imagined: size dimorphism shows a strong relationship with the opportunity for sexual selection, a relationship that holds good even when body size and phylogeny are controlled. The hypothesis that large size in males is a sexually-selected adaptation also receives some support from data on sexual selection in progress (chapter 7), which show a weak trend toward large males being more successful than small. We again stress, however, that we do not regard the comparative and sexual-selection-in-progress tests to be as conclusive as experimental ones.

Finally, the evidence is not conclusive on aggression either. An intrasexual function of aggression is supported by the fact that the most extreme forms of intraspecific aggression are used only in defense of territory. Testosterone treatment changes aggressiveness in captive males, but its effects on success in territory competition are uncertain. Combinations of testosterone blockers seem to lower success in competition for territory, but their effects on aggression are questionable. Clearly, more experimental work is needed on the effects of male aggressiveness in field situations.

9 Polygyny, Sexual Selection, and Female Red-winged Blackbirds

In the last two chapters we discussed the effects of polygyny on male red-winged blackbirds in terms of the intensified sexual selection that polygyny imposes on males. We now turn to effects of polygyny on female red-winged blackbirds. The key feature of territorial polygyny, as a social system, is that multiple females breed on the territory of a single male. This places females in close proximity to one another for several weeks, in a situation in which they are almost forced to interact in one way or another. Female redwings do in fact interact frequently with fellow harem-members, mostly in ways that seem competitive, though some mutualism may be involved also. A principal focus in this chapter will be on examining these female-female interactions within harems.

Another way in which polygyny affects females is by complicating their relationships with males. Male red-winged blackbirds, instead of each bonding to a single female, often associate simultaneously with several females, sometimes as many as 15 or more. As a consequence, males are likely to be less familiar with each of their associated females, and the bond between male and female is likely to be considerably looser, than in a monogamous species. This bond is important to the female, in that the male controls resources to which he can allow or deny her access, specifically his own parental care and the resources contained on the territory. Furthermore, the larger and more aggressive males are capable of harassing females to a considerable degree, if they so choose. A second focus of this chapter is to examine the relationships between females and males and the ways in which females try to influence these relationships in their own favor.

Finally, we want to discuss the female side of those traits we have discussed as sexually-selected adaptations in males: visual and vocal display, size, and aggressiveness. As we have seen, there is evidence that sexual selection has acted to exaggerate or enhance each of these traits in males; therefore the evolution of sexual dimorphism in these traits can be partially ascribed to sexual selection acting on miles. To understand sexual dimorphism fully, however, we must also understand how these traits have been shaped in females, and in fact some explanations of sexual dimorphism put more emphasis on selection acting on females than on males. Here we will examine what is known about selection acting on visual and vocal display, size, and aggressiveness in female red-winged blackbirds.

The Social Organization of Harems

USE OF SPACE BY FEMALES

In chapter 3 we examined the spacing behavior of male red-winged blackbirds during the breeding season and saw that male redwings exhibit classic territoriality. Male redwings satisfy both behavioral definitions of territoriality, which specify that individuals should show space-centered advertisement and aggression, and ecological definitions, which specify that individuals should have exclusive access to an area and the resources it contains. In this section we examine use of space within male territories by female redwings, in order to test ecological definitions of territoriality for females; in the next section we address behavioral definitions.

It has often been claimed that female red-winged blackbirds divide up the territories of their mates into smaller "subterritories," which they defend against other members of the harem as well as against outside females (e.g., Nero 1956a, b; Orians 1961; Wiens 1965). This claim was originally based on unsystematic observations of unmarked females. The first quantitative investigation of female spacing behavior using color-marked females was that of Hurly and Robertson (1984). These authors placed a 10- × -10 meter grid of poles in a small cattail marsh in Ontario and mapped locations of females relative to this grid for the territories of four males in each of two years. Figure 9.1 illustrates those locations. For the five territories occupied by more than one individually-recognizable female, locations are included only for periods when two or more females simultaneously resided on the territory. For three of these territories (A, E, and H), there was little overlap in female positions, whereas for the remaining two (D and G) there was considerable overlap. Hurly and Robertson report that much of the overlap in the latter two territories was due to aggressive chases during the first few days when the second female was entering the territory. They conclude that females utilize fairly exclusive areas and are in that sense territorial.

A second quantitative study of female spacing was done by Searcy (1986) on a cattail marsh in Pennsylvania. Lines of poles were set out at five-meter intervals to form a coordinate system encompassing the territories of three males in one year. These territories were eventually occupied by six, three,

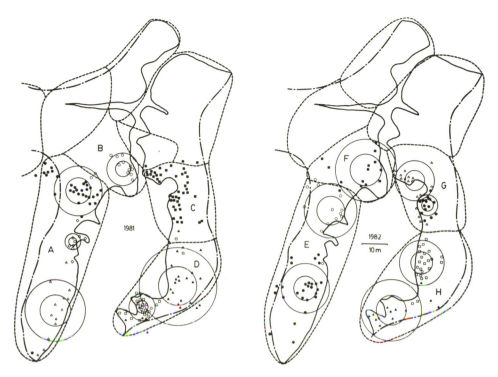

Figure 9.1 Female use of space on a small marsh in Ontario during each of two years, from Hurly and Robertson (1984). Dashed lines delimit male territories; solid concentric circles show the median and twice-the-median deviations from centers of activity for individual females. Each symbol indicates a single sighting of a female. Two identifiable females were present with little observed overlap on territories E (1982) and H (1982), and three on territory A (1981). Greater overlap between females occurred on territories D (1981) and G (1982).

and two females, respectively. The basic method of observation was to follow focal individuals for 15-minute periods, noting their locations in terms of 5 × 5-meter squares every 30 seconds. Observations were made throughout the breeding season, and in this way large samples of locations were accumulated for each female (142 to 1835 per individual). A niche overlap index (Pianka 1975) was used to calculate overlap in use of space between pairs of females occupying the same male territory (see Waser and Wiley 1979); overlap values can vary from 0 (no overlap) to 1 (completely congruent distributions). Observed overlap values were compared with "random overlaps," calculated as the overlap between a distribution observed for a certain individual and a distribution of randomly assigned points drawn from the same male's territory. The observed female-female overlaps were on average quite small (mean = 0.075; range 0.004–0.358), and a large proportion (29 of 38) were significantly lower than random overlaps.

The low female-female overlaps are consistent with female territoriality, i.e., they might be due to females repulsing each other from defended areas. A simpler explanation for low overlap is that females spend much of their time at their own nests; such behavior would bias distributions toward low overlap as long as nest sites happened to be on separate grid squares. As there were more than 100 grid squares on each of the three male territories, the chances of two females nesting on the same square would have been low, and indeed this did not occur. To test this alternative explanation of low overlap, female-female overlaps were recalculated with observations eliminated for grid squares containing nests. This analysis was restricted to females for which there were more than 100 observations after nest sites were eliminated. The resulting female-female overlaps were considerably higher (mean = 0.360, range = 0.111–0.503) than in the original analysis, and fully half (10 of 20) were significantly larger than random overlaps, whereas only one-quarter (five of 20) were significantly smaller.

One problem with the above analysis is that observations were included for the earlier-settling member of each pair of females for the period before the second female settled. Distributions might be more exclusive if we consider only the period after both females had settled, as it may be that the earlier female is forced to contract her use of space by the settlement of the second female. To test this possibility, overlaps were recalculated with nest sites eliminated for each pair, using only data taken after both females had settled (Table 9.1). The number of pairs for which more than 100 locations were available was smaller, but the average overlap (0.360) was unchanged from the previous analysis. Eight of the observed overlaps were significantly larger than random overlaps, and none were significantly smaller.

The last two analyses demonstrate that at least in one population, female red-winged blackbirds overlap in use of space more than expected by chance, when observations at their nest sites are eliminated. The fact that high overlap is only found when nest sites are included raises the question, are nest sites themselves spread out more than expected by chance? Data to assess this question were gathered by Yasukawa et al. (1992a) for an upland population in Wisconsin. These authors measured nearest-neighbor distances between nests and compared them to expected distances under a random distribution, calculated using the method of Sinclair (1985), which corrects for edge effects. Nearest-neighbor distances between primary females on different male territories were greater than expected (Table 9.2), which supports our previous conclusion that male redwings are territorial. When all simultaneously active nests are included (whether or not they are on the same male territory), however, observed nearest-neighbor distances are either equal to expected (two years) or significantly below expected (one year) (Table 9.2). Thus, this analysis suggests that nest sites are either randomly placed or clumped.

Table 9.1
Spatial Overlaps between Pairs of Females Residing on the Same Male Territories in a
Pennsylvania Marsh.

	Overlap On			
Overlap By	f1A	f1B	f1D	f1F
f1A	—	0.465** (655)	0.405* (299)	0.285 (126)
f1B	0.465** (923)	—	0.266 (299)	x
f1D	0.405** (324)	0.266 (285)	—	x
f1F	0.285 (106)	x[a]	x	—

	Overlap On		
Overlap By	f2A	f2B	f2C
f2A	—	0.309 (836)	0.468** (203)
f2B	0.309 (397)	—	0.312* (203)
f2C	0.468** (218)	0.312 (481)	—

	Overlap On	
Overlap By	f4A	f4B
f4A	—	0.371 (428)
f4B	0.371** (474)	—

SOURCE: Searcy (1986).
NOTE: Observations at nest sites have been eliminated, as have observations of one female before
the other in the dyad had settled. Sample sizes for individuals above are in parentheses.
[a]x indicates insufficient observations (<100) available for overlap calculations.
*Observed overlap was significantly greater than random at P < 0.05.
**Observed overlap was significantly greater than random at P < 0.01.

Table 9.2
Spacing of Nests between and within Male Territories in an Upland Population of
Red-winged Blackbirds in Wisconsin.

	Year	N	Observed	Expected	P
Between Male Territories	1984	10	56.1 ± 3.4	37.4	0.05
	1985	18	54.3 ± 3.2	36.1	0.01
	1986	18	45.9 ± 4.9	36.0	0.05
Within Male Territories	1984	14	29.2 ± 5.2	29.2	NS
	1985	26	30.8 ± 4.4	29.4	NS
	1985	37	19.1 ± 1.3	24.3	0.05

SOURCE: Yasukawa et al. (1992a).
NOTE: Spacing "between territories" gives mean (± SE) nearest-neighbor distances for primary
nests on different male territories. Spacing "within territories" gives mean (± SE) nearest-neighbor distances for simultaneously active nests within male territories. Observed values are compared to expected values calculated by the method of Sinclair (1985).

Of the three quantitative studies of female use of space reviewed above, one supports female territoriality (Hurly and Robertson 1984) and two do not (Searcy 1986, Yasukawa et al. 1992a). Although we cannot pretend to be unbiased, we believe the evidence in the latter two studies is stronger, especially because of the larger sample sizes of female locations in Searcy (1986) compared to Hurly and Robertson (1984). It may be unnecessary, however, to reject one result in favor of the others; rather, it may be that females use relatively exclusive areas in some populations (e.g., in Ontario) but not in others (e.g., Pennsylvania and Wisconsin).

SPACE-CENTERED AGGRESSION AND ADVERTISEMENT

Behavioral definitions of territoriality specify that individuals must show aggression and advertisement that is space centered, in the sense that each individual performs higher levels of aggression and advertisement within a certain area (the territory) than elsewhere. It has long been known that female red-winged blackbirds are at times aggressive toward one another during the breeding season, attacking and chasing one another, sometimes quite vigorously (Nero 1956b). It is also known that female redwings will attack taxidermic mounts of other females placed near their nests (LaPrade and Graves 1982, Yasukawa and Searcy 1982) and that females produce visual and vocal displays that can be interpreted as advertisement (Nero 1956a, b; Orians and Christman 1968). What remains to be seen is whether these aggressive and advertisement behaviors are space centered.

A simple test of whether aggression is space centered was carried out by Searcy (1986). In a series of trials, a taxidermic mount or a caged female was presented somewhere along the line connecting the nests of two females nesting on the same male's territory. The stimulus was always placed substantially nearer to one nest than the other. An observer then scored the aggressive reaction of both females on a five-point scale, from no reaction, through various levels of approach, up to direct physical contact with the stimulus. When two females in one trial tied for the highest criterion met, latencies were used to break the tie. In 11 trials, the female with the closer nest was more aggressive, compared to only two trials in which the farther female was more aggressive ($\chi^2 = 6.2$, P < 0.05). This experiment, then, supports the existence of at least a crude spatial gradient in aggressiveness.

Searcy (1986) also tested whether naturally occurring aggression and advertisement were spatially restricted to exclusive areas. For seven females, four or more aggressive acts were observed in which the focal female either attacked another or performed the bill-up display. The mean number of aggressive acts recorded for these seven females was 14 (range = 4–26). "Areas defended by aggression" were delimited for each of these females by drawing least convex polygons around the locations of the aggressive acts. Overlaps were calculated for pairs of females residing on the same male's territory, as the shared area divided by the area of the overlapped territory. Overlap was high for all pairs of females, with a mean of 0.67 and a range of 0.43 to 0.88. These female-female overlaps can be compared to the overlap between "areas defended by aggression" for one pair of neighboring males, which was an order of magnitude lower (0.07).

Searcy (1986) also used locations of song perches to delimit "areas defended by advertisement." Perches where females gave either of the two main categories of song, or intermediates (see below), were included. Large samples of song observations were available for some females, so it was possible to limit the analysis for each pair of females to observations made after both members of the pair had settled (this was not done for the observations of aggression). Figure 9.2 shows the locations of songs for one pair of females, f1A and f1B, together with the least convex polygons that define the areas defended by advertisement. In this pair, the overlap of A on B is 0.82, and the overlap of B on A is 0.67. The average overlap between six pairs of females was 0.65 (range = 0.34–0.93). For comparison, the overlap in areas defended by advertisement for three pairs of neighboring males

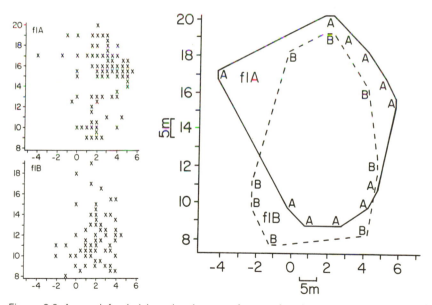

Figure 9.2 Areas defended by advertisement for two females occupying the same male's territory on a marsh in Pennsylvania, from Searcy (1986). At left are shown the locations of those grid squares on which each female was observed to sing. At right are shown the least convex polygons defined by these points. The overlap of A on B is 0.82; the overlap of B on A is 0.67.

averaged 0.06 (range = 0.03–0.09), again an order of magnitude lower than for females (see chapter 3).

The evidence above shows that female-female aggression during the breeding season is space centered, in that each female tends to be more aggressive near her nest than farther away. In this respect, then, female redwings meet the criterion for territoriality posed by behavioral definitions. However, the areas defended by females through aggression and advertisement are not at all exclusive, especially when compared to the areas defended by male redwings.

DOMINANCE WITHIN HAREMS

Before we state our conclusions on whether female red-winged blackbirds should be termed territorial, we need to present what is known about dominance relationships between females within harems. Searcy (1986) observed 15 interactions between pairs of females resident in the same male territories, for which both the winner and loser could be identified. One trend in these data was that in any pair of interactants, the female who had settled earlier tended to win; this rule ("earlier settler wins") was obeyed in 12 encounters and broken in three ($\chi^2 = 5.5$, P < 0.05). A second rule that was usually followed, but which was not statistically significant, was that the female closer to her nest tended to win; this rule ("closer female wins") was obeyed in nine of the 13 encounters in which the location could be specified as being closer to one nest than the other ($\chi^2 = 1.9$, P > 0.05). Residents were more likely to show tolerance toward each other than toward nonresidents. Resident females were almost universally dominant over nonresidents; residents attacked and chased nonresidents on 57 occasions, whereas nonresidents attacked and chased residents only twice.

Further observations on dominance relationships were carried out by Roberts and Searcy (1988), who used feeding platforms baited with grain to increase the frequency and visibility of encounters. Observations were made on three territories, containing five, five, and four females. A total of 460 encounters were observed. The earlier settler won the majority of encounters in 19 of 23 dyads, significantly more than expected by chance ($\chi^2 = 9.78$, P < 0.01). The female whose nest was closer to the platform won the majority of encounters in 30 of 45 dyads, which is also significantly more than expected by chance ($\chi^2 = 5.0$, P < 0.05). The best evidence for position effects came from experiments in which the feeding platform was moved. For example, in one such experiment the platform was initially placed close to the nest of female BO and far from the nest of female GO, and here BO won 15 of 15 encounters (100%) with GO. The platform was then moved to a position close to the nest of GO, and here BO won only two of 13 encounters (15%) with GO. Moving the platform caused significant reversals in dominance such as this in four of four dyads tested. A partial correlation analysis showed that length of residency and proximity to the nest were both significantly related to dominance independently of each other. Age, size, and stage of the nesting cycle did not show consistent relationships with dominance.

In summary, female red-winged blackbirds meet some of the criteria for territoriality but do not meet others. On the positive side, female redwings show lower than expected overlap in use of space when all spatial observations are included. In addition, they show spatial variation in aggression and dominance, being more aggressive and more dominant near their nests than farther away. On the negative side, females show higher than expected overlap in use of space when nest sites are eliminated, suggesting that any spatial segregation within male territories is due only to females concentrating their activity at their own nests. Nest sites are not spaced more widely than expected by chance. Again on the negative side, areas defended by aggression and advertisement overlap widely for females, much more widely than for males.

A final decision on whether female red-winged blackbirds are territorial depends on which of the above criteria one thinks are important. Personally, we would prefer to restrict the term "territorial" to cases in which both the behavioral criteria of spatially restricted aggression and dominance and the ecological criterion of low overlap in use of space are met (see also Wilson 1975, Brown 1975). Therefore, we would not label the areas used by female red-winged blackbirds as "territories" but would instead call them "dominions." Dominions are defined by Brown (1975) as "areas of dominance from which submissive individuals are *not* excluded," characterized by "the concurrence of aggressive behavior, uniform dispersion, and a moderate to large overlap in home range." Female red-winged blackbirds meet these specifications well, except that it is not clear that there is any real tendency toward uniform dispersion.

Female Plumage

The body plumage of females is dark brown above and pale below with dark brown streaks. Females, like males, have epaulets at the bend of the wing, though these are duller and smaller than in adult males. Epaulet size and coloration are extremely variable in females; epaulet color, for example, varies from pale yellow to red-orange. Females are also variable in chin coloration, which ranges between pale yellow and yellowish pink (Muma and Weatherhead 1991). Variation in plumage coloration is, at least in part, age related. Payne (1969) examined known-age museum specimens and found that females with reddish epaulets were usually at least two years old, and Crawford (1977a) found in the field that older females have brighter epaulets and chins than do yearling females. In individual females captured in successive springs, epaulet color becomes brighter between years in most cases (Muma and Weatherhead 1989, Torgier S. Johnsen pers. comm.), as it does in captive females (Miskimen 1980). In an unpublished study, Torgier S. Johnsen and colleagues (pers. comm.) provided evidence that changes in epaulet color between years are condition dependent. For example, these researchers showed that females that put forth a high reproductive effort in one year are less likely to increase in epaulet brightness than are females experiencing lower effort, and that increases in brightness are more likely following a year of high food abundance.

As in males, females expose their epaulets in song-spread displays, especially during aggressive, intrasexual interactions (Nero 1956a, Orians and Christman 1968). This pattern of use suggests that epaulets function as an aggressive signal. The fact that female plumage is age dependent and perhaps condition dependent suggests that aspects of the plumage may serve as status signals (Rohwer 1975). Tests of these ideas were carried out by Muma and Weatherhead (1989, 1991). In one experiment, Muma and Weatherhead (1991) formed eight captive flocks of between six and 11 females during the early spring. Epaulet length and chin length were measured as the farthest extent of coloration downward from the bend of the wing and the base of the bill, respectively. Epaulet and chin colors were scored by matching them to standard color swatches (Smithe 1975), producing a ranking that correlated well with the chroma (or intensity) of the color according to the Munsell (1961) scheme of color notation. Dominance relations were evaluated on the basis of at least 10 aggressive interactions per dyad. Considering all five groups together, there was no significant tendency for the dominant individual within each dyad to have larger or brighter epaulets or chins (Table 9.3).

In further experiments, Muma and Weatherhead (1991) manipulated epaulet and chin color by bleaching the feathers and then recoloring using a magic marker. These manipulations had no consistent effects on dominance. For example, in one experiment nine females with dull epaulets were observed for two weeks to determine premanipulation dominance ranks. Then four of these were given bright epaulets. Of these four, two remained at the same rank, one went up in rank, and one went down. Effects of manipulating chin color were similarly negative.

Muma and Weatherhead (1989) also looked for effects of female coloration on interactions in the field. In one experiment they assessed female aggressiveness toward taxidermic mounts rated as having either bright or dull chins and epaulets. Of 38 females tested with both bright and dull mounts, eight attacked the bright and five the dull; these proportions are obviously not statistically different. Furthermore, in these and other experi-

Table 9.3

Relationship between Dominance and the Size and Brightness of Color Patches of Female Red-winged Blackbirds.

Trait	N^a	Percent of Dyads in Which Female Was Dominant	P
Longer Epaulets	228	47.8	NS
Brighter Epaulets	201	47.8	NS
Longer Chin	227	48.0	NS
Brighter Chin	171	53.2	NS

SOURCE: Muma and Weatherhead (1991).
NOTE: Dominance was assessed within captive flocks. Females with longer or brighter epaulets and chins were not dominant within dyads more often than expected by chance.
[a]N is the number of dyads.

ments, a female's own plumage brightness did not predict her aggressiveness toward a mount. Finally, dull and bright female mounts were presented simultaneously to each of 18 territorial males. The males spent nearly equal amounts of time close to the bright and dull mounts, and five attempted copulation with the bright, compared to four with the dull.

In conclusion, there is no evidence that female plumage coloration has any effect on either intrasexual or intersexual interactions, and thus no evidence of a selective advantage of bright plumage. Muma and Weatherhead (1989, 1991) suggest that bright epaulets, at least, may occur in females only due to a genetic correlation with males. In other words, perhaps selection has favored epaulet coloration in males only, and female epaulets have evolved simply because plumage coloration is controlled by the same genes in the two sexes, and bright, successful males pass genes for brightness to both sons and daughters. This is an attractive hypothesis, but it should not dissuade us from continuing to search for selective consequences to females of their coloration. Manipulation of female epaulet size and color in the field would be particularly interesting.

Female Vocal Behavior

TYPES OF VOCALIZATIONS

The simpler vocalizations produced by female red-winged blackbirds are termed "calls." Females and males share some clicklike "check" calls (Nero 1956a, Orians and Christman 1968), and these seem to function as contact or mild alarm calls. Some other calls, such as the "scream," are given only by females. Screams are produced during intense alarm situations, as when a predator discovers an active nest. Screams function to attract other redwings to the vicinity of the nest and also cause nestlings to behave more cryptically (Knight and Temple 1988; see chapter 3).

The longest, most complex vocalizations produced by female red-winged blackbirds are termed "songs." Female songs sound quite unlike male songs. Furthermore, the songs of female redwings cannot be divided into a discrete number of distinct song types in the way that male songs can. Instead, females produce three or more categories of elements, which they combine in many different ways. The resulting complexity makes the analysis and classification of female songs a difficult task.

Nero (1956a) described female red-winged blackbird song as "a series of high, shrill and rapid notes, slowing and descending at the end, the last phrase often very sibilant and slurred" and "given with considerable variation." Nero renders female song into words as "spit-a-chew-chew-chew" or "check-check-a-skew-skew." Orians and Christman (1968) also describe female song as highly variable and render it as "Ch-ch-ch-che-chee-cheee" or "Cheeee-cheeee-cheeee." These authors are clearly describing the same "female chatter" vocalization. An inspection of the sound spectrograms provided in Orians and Christman (1968, Fig. 15) shows them to contain at least two major vocal elements, the "chit" ("check-check-a" or "ch-ch-ch") and the "teer" ("chew-chew-chew" or "cheeee-cheeee-cheeee"). The variability noted by both sets of authors is in part caused by individual differences but also in part occurs within individuals.

More recent analyses have attempted to classify the variable songs of female redwings into a limited number of categories. Dickinson (1987), after extensive spectrographic analysis, concluded that female redwing song consists of three elements, which he termed "chit," "teer," and "chet." Armstrong (1992) used visual classification of spectrograms and multivariate statistical analysis of structural components to identify five elements in female song: Dickinson's "chit," "teer," and "check" (= "chet"), and in addition "hee" and "ti" (Figure 9.3). Examination of spectrograms of many examples of each element shows some of them to represent points along a continuum. Dickinson (1987), for example, notes that chits grade into chets, and Armstrong (1992) found grading between check (chet) and chit, and between chit and hee (Figure 9.3).

Although sophisticated analysis of spectrograms demonstrates the complexity of song elements produced by females, it is very difficult for human listeners to distinguish many of the elements (Armstrong 1992). For this reason, and to increase interobserver reliability, many of us who study the vocal behavior of female red-winged blackbirds have adopted a simpler classification system, in which certain of the categories described above are collapsed into each other. Three categories are recognized in this scheme: (1) chits (Hurly and Robertson 1984), which are songs containing chit, chet, hee, and/or ti; these were called "type 1" songs by Beletsky (1983a); (2) teers (Hurly and Robertson 1984), which are songs containing only teer

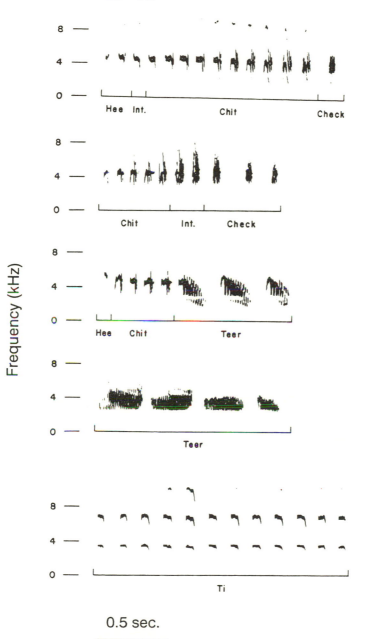

0.5 sec.

Figure 9.3 Spectrograms of the chief elements making up the songs of female red-winged blackbirds, from Armstrong (1992). Note that besides the hee, chit, check, teer, and ti, there are also illustrated intermediates between hee and chit and between chit and check.

elements; these were called type 2 songs by Beletsky (1983a); and (3) composites (Beletsky 1983a, Armstrong 1992), which are songs containing teers combined with other elements. In a large sample of songs, Dickinson (1987) found that 70% were chits, 5% were teers, and 25% were composites.

CONTEXTS OF SONGS

Teer songs are associated with aggressive, intrasexual interactions (Beletsky 1983a). For example, when we presented stuffed mounts of female red-winged blackbirds to nesting females in an attempt to measure their aggressive response, nesting females frequently responded with teer songs before attacking the mount (Yasukawa and Searcy 1982). Yasukawa (unpublished) observed similar responses by resident females to a live, caged female. Beletsky (1983b) showed that rates of teers increased dramatically in resident females when he simulated intrusion with playback of either teers or chits. Testosterone implants, which in male redwings increase aggressive behaviors (Searcy and Wingfield 1980), greatly increased rates of teers in free-living females (Searcy 1988a). Resident females will also occasionally sing teers in response to strange males that fly over or perch nearby.

Chit songs are associated primarily with intersexual and parent-offspring interactions. Chits are often sung in response to the song of the male (Beletsky 1985). Such intersexual song answering is a striking feature of female singing. A female is able to respond to a male song quickly enough that her song may overlap up to 80% of the male song. This overlapping led Beletsky (1983a) to refer to such performances as duetting. In one sample of 3727 male songs given when females were present on the territory, females answered 1710 (46%) (Beletsky 1985). Almost all of the answers (94%) were chits. Females also produce chits frequently when in close proximity to their active nests. This "nest-associated" singing (Beletsky and Orians 1985) occurs as the female arrives at her nest, as she leaves her nest, and even as she incubates eggs or broods young. Nest-associated songs virtually always contain chits and chets, but on occasion females add teers, producing composite songs (Beletsky and Orians 1985, Yasukawa 1989).

Another aspect of the context in which teers and chits are given is the seasonal pattern of their use. Yasukawa et al. (1987a) lumped composite songs with teers and examined seasonal effects by date and by stage of the nesting cycle. Teer rates were found to be highest (about 1/minute on average) during the prenesting period (late April) and low (less than 0.4/minute) once nesting began. This result again suggests that teers are aggressive, as female-female aggressive encounters are most common before nesting begins. In contrast, chits were sung at relatively constant rates (more than 0.5/minute) throughout the breeding season and nesting cycle, except that females rarely sang at all once their young fledged. Other studies have shown similar seasonal patterns (Beletsky and Orians 1985, Small and Boersma 1990).

FUNCTION OF TEERS

The information presented above on the context of teers strongly supports an aggressive, threatening function for this vocalization, as Beletsky (1983a) originally hypothesized. In sum, this evidence shows that female red-winged blackbirds give teers mainly in situations in which they are competing aggressively with other, conspecific females. The most direct experimental test of this function has given rather equivocal results, however. This was a speaker-occupation experiment performed by Yasukawa (1990), which we described in some detail in chapter 5. In brief, Yasukawa chose matched pairs of male territories and placed a cassette player and two speakers in a randomly chosen territory in each pair. Teers were broadcast from the speakers on experimental territories for five hours each morning through the period of settlement by primary females. In 11 of 14 cases, the experimental territory in a pair (the one with the teer playback) was settled before the control. Overall, dates of settlement were significantly earlier on the experimental territories than on controls. Thus, teers seem to encourage, rather than discourage, settlement by other females.

Taken at face value, the speaker-occupation results suggest that teers have a cooperative, attractive function rather than an aggressive, threatening one. Nevertheless, we persist in believing the contextual evidence in favor of the aggressive function. Teers seem to us to be too closely tied to aggressive contexts to admit any other function. Particularly convincing is the association of teers with direct, physical attack on female stimuli (such as mounts or caged females), in which the aggressive nature of the context is undeniable. What the speaker-occupation results say to us is that female aggressive behavior, including teers, does not function in deterring settlement. We will consider alternative functions for female aggression in general later in the chapter.

FUNCTION OF CHITS

One of the striking features of the contexts of chits is that they are so often given by females when leaving and entering their nests. Such nest-associated vocalizations would be expected to have some cost, in making the nest more conspicuous to predators. This cost was confirmed by Yasukawa (1989). Old redwing nests, containing plaster eggs, were placed on deserted redwing territories, and speakers were placed at the base of the nest supports. From half of the speakers chits were played, from sunrise to sunset, at the rate of one per minute. Nests with playback were depredated significantly sooner than nests without playback; most of the predation was apparently by mink.

We would expect nest-associated chits to have some especially strong benefits to compensate for the cost of making the nest more conspicuous. Beletsky and Orians (1985) suggested two possible compensating benefits: (1) in coordinating movements of females and their male mates, so that the nest is not left unguarded, and (2) in coordinating movements of neighboring females, allowing them to leave their nests simultaneously, which may provide increased safety from predation. These authors found, however, that when male territory owners were removed from territories, their females nevertheless vocalized during 80% of their nest departures, which does not accord well with the male-female coordination hypothesis. Further, they found that most females depart singly from the territory rather than in groups and that playback of chits did not stimulate nest-leaving by nearby females. These latter findings do not accord well with the female-female coordination hypothesis.

Following the rejection of their initial hypotheses on the function of chit vocalizations, Beletsky and Orians (1985) proposed a new hypothesis, that a female gives chits in order to inform her mate about her activities and reproductive state, thus encouraging the male to provide parental help at the appropriate stages of the nesting cycle, and discouraging him from harassing the female, as males sometimes harass nonharem females. The experiments of Yasukawa (1989) strongly support this new hypothesis. Yasukawa set out dummy nests, which contained plaster eggs, that mimicked redwing nests, from some of which he played chit vocalizations, while others served as silent controls. When he subsequently tested male response to crow models placed at these nests, males defended the nests from which chits had been played significantly more strongly than control nests. Chit songs, then, do seem to encourage parental help from the male.

There is also evidence that chits discourage harassment by males. Birks and Beletsky (1987) observed arrivals and departures of female red-winged blackbirds from their nests and noted whether or not the females vocalized and whether or not they were chased by their mates. Out of the 70 occasions when females vocalized during arrival or departure, they were chased only once (1.4%), whereas out of the 21 occasions when they arrived or departed silently, they were chased eight times (38.1%). We conclude that both benefits proposed by Beletsky and Orians (1985) are operating: chits encourage parental help and discourage harassment from the female's mate.

Another striking aspect of the context of chit songs is that they are given so often in reply to male song. Females reply in this fashion both when on the nest and when elsewhere in the territory. If replying to the male functions to reinforce the pair bond, making paternal care more likely and harassment less likely, and if replying has some cost (such as increased conspicuousness to predators), then we might expect it would be advantageous for a female to learn to recognize the songs of her mate and reply only to those. Beletsky and Corral (1983) conducted a small study of vocal mate recognition by playing mate and neighboring male songs to each of three females. These females replied to 43% of mate songs and 41% of nonmate songs, thus providing no evidence of mate recognition. In a larger study, Yasukawa (1989) played three categories of male songs—mate, neighbor, and nonneighbor (recorded five kilometers away)—to females as they sat on their nests incubating their eggs. A total of 48 females were tested in the absence of their mates, and each heard a single example of each type of male song in a random order. Females responded to 43.6% of mate songs, 31.1% of neighbor songs, and 44.9% of nonneighbor songs. Again, these results offer no evidence of mate recognition. Nevertheless, in natural inter-actions females usually reply only to their mates. We believe that a simpler rule—reply to loud, nearby male songs—produces a specific enough system of song answering, without females being able to discriminate mate from nonmate song by their acoustic properties. Given the frequency with which female redwings change mates and the existence of male song repertoires, such a system may be the best one possible.

Evidence that song answering is of overall benefit to female redwings comes from a comparison of answering rates in females that nested suc-cessfully (produced at least one fledgling) or unsuccessfully (produced no fledglings) (Yasukawa 1989). Because primary females were significantly more successful than nonprimary females, the analysis was done separately for these two classes. Among primary females, chit answer rates (answers/hour) were significantly higher among successful females than among unsuc-cessful (Figure 9.4). The same relationship held true for nonprimary females as well (Figure 9.4).

Female Size and Size Dimorphism

MODELING SIZE DIMORPHISM

We have seen that males are larger than females in icterines and that the degree of dimorphism is greater in polygynous than monogamous species. Such a pattern is of course very common in birds (Selander 1972) and mam-mals (Clutton-Brock et al. 1977, Alexander et al. 1979), though it is not so common in other animals (Ghiselin 1974). Both aspects of the pattern can be ascribed to sexual selection: sexual selection favors large size in males and acts more powerfully in polygynous than monogamous species. This expla-nation, however, leaves out any account of how selection acts on size in females. Could selection for smaller size in females cause the "normal" pat-

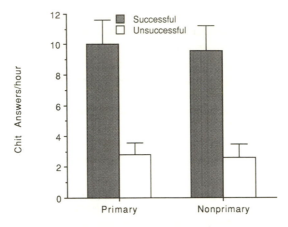

Figure 9.4 Rates of chit answering by females successful and unsuccessful in nesting, from Yasukawa (1989). Chit answering is a behavior in which females answer a male song with one of their own chit songs. Successful nesters are females producing at least one fledgling. Successful nesters gave significantly higher rates of chit answers than unsuccessful nesters for both primary females (P < 0.05) and nonprimary females (P < 0.01).

tern of size dimorphism in the higher vertebrates, rather than selection for larger size in males? In the absence of sexual selection, would males be the same size as females, or are there other selection pressures acting differentially on the two sexes?

Downhower (1976) has suggested that energetic considerations alone may lead to the evolution of smaller size in female birds than in males. Downhower pointed out that metabolic rate increases as an exponential function of body mass, i.e., $R = aM^b$, where R equals metabolic rate, M stands for body mass, and a and b are constants. For standard metabolic rate (i.e., of a resting animal in the thermoneutral zone), the exponent b is found to equal 0.72 in between-species comparisons of passerine birds (Calder 1984). Fat stores also increase allometrically with size, i.e., $M_{fat} = cM^d$, where M_{fat} is the mass of the fat stores, and c and d are constants. The exponent d in the latter equation is thought to be larger than 0.72; Calder (1984) assumes a value of approximately 1.0. It follows that fasting endurance, the time an individual is able to survive on its fat stores alone, increases with body mass. Downhower (1976) argued that male birds, especially in polygynous species, arrive at the breeding grounds earlier than females so that they can acquire a territory. During this late winter/early spring period, weather is poor and males must spend much of their time in territory defense rather than in foraging. Therefore, males ought to be selected for fasting endurance so that they can support territory defense on their fat reserves. Downhower argued that females, in contrast, produce eggs out of immediate energy surpluses rather than from long-term storage. Small females ought to have lower maintenance requirements, and thus will start generating the surpluses necessary for egg formation earlier in the season. Early breeding is usually advantageous, so small size will be favored in females.

Downhower's argument rests on the assumption that it is easier to meet the energy costs of territory defense out of energy stores than it is to meet the costs of egg formation. This assumption seems doubtful, at least when applied to red-winged blackbirds. Ricklefs (1974) stated that the daily energy cost of egg formation in altricial passerines is about 45% of basal metabolic rate (BMR), which would typically be expended over a period of four or five days in redwings. Orians (1961) estimates the daily energy costs of territory defense during this period as approximately 11% of BMR. Therefore, the total energy expenditure of males in territory defense prior to nesting would be higher than female expenditure on egg formation as long as males set up territory 16 to 20 days before females begin nesting. In fact, male redwings typically defend territories 30 to 90 days before nesting begins. Thus, fat stores would make a larger contribution to egg formation than territory defense if body sizes were equal in the sexes.

Downhower's (1976) "dual energetic optima" hypothesis suggested that the optimum size from the standpoint of energetics is different in females than males. By contrast, Searcy (1979c) suggested that the energetic optimum ought usually to be the same in the sexes within a species. The argument assumes that the feeding niches of the sexes are similar enough that a single function, $f(s)$, describes the way in which the energy profit of foraging maps on size (s). If we let $g(s)$ represent the maintenance costs of the organism, also a function of body size, then $f(s) - g(s)$ would represent the energy left over after maintenance to expend on reproduction or activity. If we let $h(s)$ stand for the energy cost of a given unit of activity or reproduction, then $[f(s) - g(s)]/h(s)$ is a reasonable quantity for selection to maximize. If $h(s)$ is the energy cost per unit of reproduction, the units in this expression are:

$$(\text{energy/time} - \text{energy/time})/(\text{energy/unit reproduction}) = \text{unit reproduction/time}.$$

Even though we are assuming that $f(s)$ and $g(s)$ are similar functions in males and females, the optimum size will still be different in the sexes if $h(s)$ is different. It turns out, however, that many crucial energy costs scale to body size with a similar exponent, approximately 0.72–0.73. For example, egg size, and presumably egg energy content, increases with mass to the 0.73 power in passerines (Ricklefs 1974). Flight costs in birds increase with the 0.73 power of mass also (Berger and Hart 1972). Maintenance

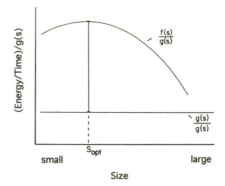

Figure 9.5 A model giving the energetically optimum size, from Searcy (1979c). f(s) is the energy profit per time foraging as a function of size (s), and g(s) is maintenance costs per time, again as a function of size. The optimum size occurs at the maximum of the function $[f(s) - g(s)]/xg(s) = 1/x[f(s)/g(s) - 1]$, where x is the factor by which maintenance costs are multiplied to give the cost of crucial activities such as reproduction or flight. Regardless of the value of x, this optimum occurs where the difference $f(s)/g(s) - 1$ is maximized, i.e., at the peak of the $f(s)/g(s)$ curve. Therefore, as long as the foraging and maintenance functions are similar in the two sexes, the energetically optimum size ought to be the same for females as for males.

costs [g(s)] can be considered to equal BMR, which increases with size to the 0.72 power. Thus the cost function h(s) can be expressed as some multiple of g(s), i.e., xg(s). In that case, the optimum size becomes that size where $[f(s) - g(s)]/xg(s) = 1/x[f(s)/g(s) - 1]$ is maximized. This is shown graphically in Figure 9.5. The value of x will vary according to what cost is most important, but this should not affect the optimum size, which will always occur at the size where f(s)/g(s) is maximal.

Searcy (1979c) hypothesized that female red-winged blackbirds would be at the energetically optimum size as given by this model and that males would be above this optimum due to the pressure of sexual selection. Langston et al. (1990) have recently suggested that female redwings will also have been driven above the energetic optimum by the pressure of social competition. This hypothesis assumes that larger females are at an advantage in social competition, for example in competition for mating opportunities. Competition for mating opportunities is sexual selection, though other kinds of social competition may be involved as well, such as competition for food within male territories. Under this "social competition" hypothesis, then, sexual selection will have favored large size in both males and females, and presumably males are larger than females because sexual selection has acted more powerfully on males.

We can make some predictions from these hypotheses that can be tested by existing data. Downhower's dual energetic optima hypothesis explicitly predicts that small females should breed earlier and consequently should have higher reproductive success than large females. Conversely, the Langston et al. social-competition hypothesis predicts that large females should breed earlier and have higher reproductive success. Searcy's "single energetic optimum" hypothesis predicts that females of median size should have the most favorable energy balance and presumably the highest survival, whereas the social-competition hypothesis predicts that small females should be at an advantage energetically and have the highest survival.

EMPIRICAL RESULTS

Less attention has been paid to female size in red-winged blackbirds than to male size, but there are some data, mainly from a study by Langston et al. (1990) on a Washington population. In this population, the probability of successful nesting decreased through the breeding season, so early nesting was advantageous as assumed by Downhower (1976). Large females tended to breed earlier than small females, the opposite of what Downhower's hypothesis predicts (Figure 9.6). The negative correlation between size and breeding date was significant for size measured as wing length, as mass, and as the first principal component of either 13 bone measures ("structural size") or of three external measures ("external size"). Langston et al. (1990) attributed the earlier breeding of larger females to success in social competi-

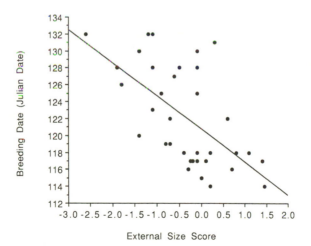

Figure 9.6 Relationship between breeding date and female size for a Washington population of red-winged blackbirds, from Langston et al. (1990). Breeding date is the Julian date on which the first egg was laid. External size score is the first principal component in an analysis of wing length, tarsus length, and mass. Larger females bred earlier in the season, when the probability of nest failure was lower. The relationship between breeding date and size is significant ($r = -0.57$, $P < 0.001$).

tion for early breeding opportunities. In support, they showed that large females were more aggressive than small females and that more aggressive females bred earlier. A problem with this interpretation is that there is not much evidence that female aggression influences settlement on territories (chapter 5). Regardless of the mechanism, it does appear that large females experience an advantage from breeding earlier, though note that the evidence is only correlative; the effect may be due to some factor other than size. Furthermore, early breeding has not always proven to be advantageous in other studies. For example, Daniel A. Cristol (pers. comm.) experimentally delayed breeding in female redwings by holding them in captivity and found that the delayed females were as likely to breed successfully as were control females, which were captured but not held.

Langston et al. (1990) also investigated the effects of size on energetic balance in females. Forty-five females were collected during breeding and classified as large or small on the basis of structural size. Lipids were extracted from their carcasses, and lipid content was estimated as the difference between dry mass and lean dry mass. Large females had significantly less fat, both on a percentage and an absolute basis, than did small females. Controlling for size, females with more fat bred earlier and laid larger eggs. These data indicate that small females are nearer to the energetically optimum size, and thus that females as a whole have been driven above that optimum.

A number of studies have investigated effects of size on survival in females, using differing methods and producing differing results. (1) Johnson et al. (1980) compared females found dead at a winter roost to those captured alive. Dead females were significantly smaller in both wing and tarsus lengths than live females (Table 9.4), indicating an advantage to being large. (2) Weatherhead et al. (1984) showed that females killed by surfactant spraying did not differ from those that survived (Table 9.4). (3) Weatherhead et al. (1987) showed that samples of yearling and adult females captured at a roost in the spring did not differ in wing lengths from same-age females captured in the roost the previous fall. (4) Langston et al. (1990) found that one-year-old females were significantly larger than older females in a principal component measure of structural size, though they did not differ in tarsus lengths (Table 9.4), indicating a survival advantage to being small. (5) Weatherhead and Clark (1994) found that museum samples of adult females captured in the fall did not differ in wing length or a principal component measure of size from those surviving to the spring (Table 9.4). (6) Weatherhead and Clark (1994) examined samples of females captured in the field, comparing those resighted in a second year to those never seen again. For both adult females and yearlings, individuals known to have survived did not differ from those never seen again in either wing length or a principal component measure of size (Table 9.4).

Table 9.4

Survival Selection on Female Body Size in Red-winged Blackbirds.

Measure	Age Class	Survivors	Controls	P	Study
Wing (mm)	All	107.8 (15)	104.5 (101)	0.01	Johnson et al. 1980
	Adults	102.5 (24)	103.9 (17)	NS	Weatherhead et al. 1987
	Yearlings	100.2 (19)	100.5 (19)	NS	
	Adults (museum)	99.7 (23)	101.9 (19)	NS	Weatherhead & Clark 1994
	Adults (field)	101.1 (20)	101.4 (109)	NS	
	Yearlings (field)	99.3 (15)	99.2 (112)	NS	
Tarsus (mm)	All	28.0 (15)	25.5 (101)	0.0001	Johnson et al. 1980
	All	30.4 (152)	30.2 (24)	NS	Langston et al. 1990
Ulna (mm)	All	28.5 (17)	28.7 (23)	—	Weatherhead et al. 1984
PCI	All	−0.1 (63)	0.6 (11)	0.05	Langston et al. 1990
	Adults (museum)	−0.16 (22)	0.43 (17)	NS	Weatherhead & Clark 1994
	Adults (field)	0.36 (20)	0.18 (93)	NS	
	Yearlings (field)	−0.20 (9)	−0.41 (44)	NS	

NOTE: Survivors are those known to have survived over a given period. Controls are either a sample of those known to have been alive at the start of the period (Weatherhead et al. 1987, Weatherhead and Clark 1994) or a sample of those known or presumed to have died during the period (Johnson et al. 1980, Weatherhead et al. 1984). Langston et al. (1990) compare adults ("survivors") with yearlings ("controls"). Sample sizes are given in parentheses.

Overall, two studies showed significant differences in size between survivors and controls, one indicating a survival advantage for large females and the other an advantage for small females. Disregarding statistical significance, trends favor large-female advantage and small-female advantage about equally often. We conclude that there is little or no directional survival selection acting on body size in female red-winged blackbirds. Weatherhead and Clark (1994) found that surviving females tended to be less variable in a principal component measure of size than those that disappeared, indicating that there may be stabilizing selection on female size.

To summarize the empirical tests of our hypotheses on female size, we can reject the dual energetic optima hypothesis of small female advantage due to early breeding, as the data indicate the opposite trend. There is nothing to contradict the assumption of our size model (Figure 9.5) that the energetic optimum is similar for males and females, but results showing that small females maintain a more favorable energy balance disprove the assumption that female size is at the energetic optimum. The social-competition hypothesis that female size has been driven above the energetic optimum is supported by the results showing that large females breed earlier and consequently have higher breeding success, and that small females are in more favorable energy balance. The social-competition hypothesis predicts, however, that small females should have a survival advantage, which does not seem to be generally true. If large females have a reproductive advantage and no survival disadvantage, it is not clear what is acting to limit female size below that of males. What is needed to clarify matters is experimental manipulation of female size, but unfortunately such experiments appear impractical at present.

Function of Female Aggression

The existence of female-female aggression is easy to observe in red-winged blackbirds, but the function of this aggression remains one of the outstanding unanswered problems in redwing behavior. We have seen bits of evidence bearing on this problem previously; here we attempt to synthesize these bits with other evidence and come to an overall conclusion. As we shall see, however, no clear answer emerges.

There are some costs to sharing the territory with other females, notably from competition for the nonshareable aspects of male parental care (chapter 6). Opposing this cost is at least one benefit to sharing the territory: increased safety of nests from predators. Nests built near other nests have higher success because of predator dilution and/or cooperative defense (chapter 4). Asking whether the costs outweigh the benefits is equivalent to asking whether polygyny has a net cost or a net benefit; we addressed this question at length in chapter 6 and concluded that costs and benefits roughly cancel, with, if anything, benefits predominating slightly.

If polygyny has no cost to females, and perhaps a slight benefit, we might expect that resident females would at least be indifferent to the settlement of other females on the territory, whereas at most they might actually encourage such settlement. Instead, females *act* as if they are trying to discourage settlement. As we have mentioned previously, resident females attack and chase nonresident females that intrude onto the male's territory (Nero 1956b). They also attack or display aggressively at mounts of females placed on the male's territory (LaPrade and Graves 1982, Searcy 1986) and approach and display aggressively in response to playback of female vocalizations (Beletsky 1983b). Moreover, certain patterns in the aggressiveness of resident females accord with the idea that the aggression is directed at preventing settlement. We suggested (Yasukawa and Searcy 1982) that if the aggression of resident females is aimed at preventing settlement by other females, then resident females ought to be more aggressive toward another female showing interest in mating with the territory owner than toward a female simply perching within the territory. To test this prediction, we placed on territories taxidermic mounts stuffed either in a normal perched posture or in a copulation solicitation posture. Of 26 females presented with the soliciting mount, 12 (46.2%) attacked, compared to only two of 21 (9.5%) that attacked the perched mount ($\chi^2 = 7.59$, P < 0.01).

It appears, then, that settlement by other females is not costly to resident females, yet residents nonetheless attempt to deter settlement. A second paradox is that even if residents are trying to deter settlement, they are not successful in doing so. We have already reviewed at length the extensive evidence on the effect of resident females on further settlement (chapter 5) and concluded that the bulk of the evidence weighs against any deterrence effect. A major reason why residents fail to discourage nonresidents from settling may be interference from the male. When a resident female attacks a nonresident, the male territory owner often intervenes by attacking the resident (Nero 1956b, Searcy 1986). Because males are larger and more aggressive than females, their intervention is quite effective; sometimes they will prevent a resident from even approaching a nonresident. In sum, what the evidence seems to say is that resident females are not able to deter settlement, and furthermore they would not benefit from deterring settlement, and yet they nevertheless seek to do so.

One way out of these paradoxes is to assume that female-female aggression is not aimed at deterring settlement after all but instead has other functions. One possibility is that female aggression functions in defense of subterritories within male territories, but we have already seen that female behavior is not consistent with territoriality, at least as we have defined it. A second possibility is that female aggression functions to establish dominance within the harem. This possibility seems more in accord with what is known. Females do establish a complex system of dominance relationships, in which relative status within a dyad may change with distance from the females' respective nests (Roberts and Searcy 1988). Order of settlement on the territory is also related to dominance, with earlier-settling females dominant to later-settling females. This last observation is consistent with the greater aggressiveness shown by primary females (Yasukawa and Searcy 1982) and suggests that the aggression shown by residents toward nonresidents functions to impose subordinate status in advance, in case the nonresidents later become residents. We also know that dominance relationships determine access to food on the territory, at least on artificial feeding platforms (Roberts and Searcy 1988), so there is at least some benefit to being dominant.

We can propose a hypothesis that accounts for most of the facts, as follows. It does not benefit resident females to deter settlement, and therefore they do not attempt to do so. Females are aggressive toward one another, with residents especially aggressive toward nonresidents, but the function of aggression is to establish dominance in order to secure priority of access to resources on the territory, and not to deter settlement. This hypothesis is plausible, but we are not entirely comfortable with it. The problem is that if one watches the behavior of a resident female early in the breeding season, doggedly chasing a prospecting female around and around the male's territory, it simply looks like that resident is trying to deter settlement. Furthermore, the fact that males intervene on the side of the prospecting female makes sense if the resident is trying to discourage settlement, but makes little sense if the females are simply fighting over dominance.

If females really are attempting to deter settlement, how can we explain this behavior? One possibility is that the behavior is nonadaptive, occurring perhaps due to a genetic correlation with male territorial aggression, or as a phylogenetic holdover from some monogamous ancestor in which female deterrence of settlement was adaptive. Another possibility is that there is some benefit of deterring settlement which we have not yet discovered, perhaps one that accrues only at certain stages of the breeding season, or that operates only sporadically between years or regions. We do not feel confident in rejecting any of these possibilities as of now; clearly, this is an area where further research is needed.

Conclusions

As a result of the red-winged blackbird's socially polygynous mating system, with many females residing in the territories of individual males, there is more of a female social system in redwings than in most species of birds during the breeding season. This female social system is not based on female territoriality. Female red-winged blackbirds residing on a male's territory are aggressive both toward other residents and toward outsiders. Female aggressiveness shows some spatial dependence, with females more aggressive close to their own nests than farther away. Female red-winged blackbirds cannot be said to be territorial, however, because they show so much overlap in use of space. Rather, females show patterns of dominance that are in part related to position in space and in part to other factors, such as order of settlement on the male's territory. Although female aggression toward nonresidents appears to be directed toward preventing further settlement on the male's territory, there is little evidence that resident females are successful in deterring settlement, nor that deterring settlement would be advantageous to residents. Aggression of residents toward nonresidents may instead function in establishing dominance relationships, or it may be maladaptive.

Female redwings have epaulets that are somewhat feeble versions of the epaulets of adult males. There is a great deal of variation among females in coloration, but so far there is little evidence that this variation is important as a social signal. Size and color of epaulets and chin displays do not correlate with dominance or aggressiveness, nor does manipulating these displays cause changes in dominance. Epaulet color in particular may occur in females only due to a genetic correlation with males; however, this is essentially a null hypothesis, and further investigation of possible benefits of visual signals in females is needed before we accept it.

Female red-winged blackbirds have a complex vocal repertoire, including several varieties of song. The latter can most easily be classified as teers, chits, and chit-teer composites. Teers are almost certainly an aggressive signal, directed primarily at other females, but the benefit of this vocalization, like the benefit of female-female aggression in general, is still in question. Chits are often given in reply to male song and are also commonly produced when a female is arriving at or departing from her nest. The function of chits seems to be to reinforce the pair bond, by keeping the male informed of the female's nesting activity in order to encourage his help, and to identify the female to the male and so discourage harassment. The singing behavior of female red-winged blackbirds, which is more elaborate and sophisticated than in most female birds, may be seen as an adaptation to polygyny, both to the greater frequency of female-female interactions and to the relative weakness of the pair bond.

Comparative evidence indicates that red-winged blackbirds have evolved size dimorphism due to sexual selection favoring larger size in males. Presumably, sexual selection has favored large size in males because large males have or had an advantage in intrasexual competition. It seems, however, that large size may also be favored in females through a similar mechanism, that is, through an advantage of large size in intrasexual competition. As Patrick J. Weatherhead (pers. comm.) has pointed out, size dimorphism should not evolve if large size is favored in both sexes, even if the advantages of large size are greater in males than in females. What this suggests is that there must be counterselection against large size, at least in females, even though attempts to demonstrate such counterselection have mainly failed so far. We propose, then, that size in females is set by a balance between social competition favoring large size and survival selection favoring small size, and that male size has been driven above this female balance by stronger selection differentials favoring large size, stemming from sexual selection. Much of this explanation remains speculative.

10 Conclusions

We have now reviewed what is known about polygyny and sexual selection in red-winged blackbirds. In doing so, we have stuck rather narrowly to our focal species, only looking at cross-species comparisons when we could use such comparisons to shed light on the evolution of the behavior and morphology of red-winged blackbirds. In this final chapter we will reverse this process and consider what light can be shed on other species by what we have learned about redwings. In other words, we will examine how well our conclusions about polygyny and sexual selection in red-winged blackbirds generalize to other species.

Polygyny

Our explanation of social polygyny in red-winged blackbirds has the following characteristics. First, polygyny results from female choice, by which we mean that females choose where to settle and with whom to mate, and thereby determine their own mating status rather than having it forced on them by males. Second, polygyny has no net cost, meaning that a female's fitness is not lowered due to one or more females having already mated with the male she chooses. Third, polygyny has no net benefit, or at least no large net benefit. Fourth, females show directed choice, meaning that their settlement is influenced by characteristics of the males and, especially, of their territories.

In this hypothesis, the key element is the lack of a cost of polygyny. Given female choice (which seems to prevail in virtually all cases of avian polygyny), the absence of a cost of polygyny makes a substantial incidence of polygyny inevitable; the other elements of our explanation, no benefit of polygyny and directed choice, then merely affect the degree of polygyny. The conclusion that polygyny has no cost in red-winged blackbirds is supported by the following evidence. (1) Females choosing already-mated males pay only a small cost due to reduced male assistance in feeding their offspring, as deduced from the benefit of male provisioning and its apportionment among females of differing status (chapters 2 and 6). (2) Females nesting on territories with other females experience a benefit due to reduced nest predation, as shown especially in experiments with artificial nests (chapter 4). (3) When harem sizes are reduced by removal of females, seasonal reproductive success of the remaining females does not increase (see chapter 6). (4) The presence of females on a territory does not discourage further settlement on that territory, at least not in most studies (chapter 5). In some ways the last type of evidence is most compelling, in that female choice should have evolved in response to any source of cost, including any sources that researchers have not yet measured separately.

WHY THERE IS NO COST OF POLYGYNY IN RED-WINGED BLACKBIRDS

One reason that polygyny has no cost in redwings is that male redwings put relatively little effort into nonshareable parental care compared to other species. The chief, perhaps the only, nonshareable form of parental care provided by male redwings is help with feeding the young. Because male redwings provide relatively little help in provisioning the young, the cost to polygynous females of sharing this help is less than it would be in most monogamous species; females losing this help are hurt less because there is less to lose. By way of illustration, Table 10.1 shows the percentage of all feeding visits (to nestlings) made by males for a sample of monogamous species. In these species, the male contribution varies from 33 to 62%, with an overall average of 49%. By contrast, in an eastern population of red-winged blackbirds with one of the highest known proportions of provisioning males (87%), and in which primary broods are favored by males, males make only 17% of feeding visits to primary nests (Yasukawa et al. 1990). The average male contribution must be considerably lower in other populations, especially in the far West where most males virtually do not feed at all. For example, in Beletsky and Orians's (1991) Washington population, only 6% of males provision; if we assume that these feeding males provision at a similar rate as feeding males in the Wisconsin population studied by

Table 10.1
Male Contribution to Feeding Nestlings in Various Species of
Primarily Monogamous Passerines.

Species	Male Contribution[a]	Reference
Great tit	52	Smith et al. 1988
Northern mockingbird	47	Breitwisch et al. 1986
Brown thrasher	58	Heagy & Best 1983
House sparrow	44	McGillivray 1984
Tree swallow	48	Leffelaar & Robertson 1986
Black-billed magpie	62	Buitron 1988
Common grackle	33	Howe 1979

[a]Contribution is measured as the percentage of feeding visits made by males.

Yasukawa et al. (1990), then Washington males would overall provide only about 1% of all feeding visits to primary nests.

Another factor that affects the cost of sharing male parental care, besides the average amount of male provisioning, is the degree of dependence of female reproductive success on male provisioning. Studies in which males of monogamous species are removed from breeding pairs have shown that male parental care is often not as important to female reproductive success as one might expect (Wolf et al. 1988, Bart and Tornes 1989). Wolf et al. (1988) concluded from a review of male-removal studies that paternal care is most likely to be important in species that nest in cavities or breed at high latitudes, neither of which applies to red-winged blackbirds. From a similar review, Bart and Tornes (1989) concluded that paternal care is very important in species in which males help incubate or bring food to females during incubation; again, neither applies to redwings. Paternal care also seems to be particularly important in species in which the young need nearly constant brooding immediately after hatching (Bart and Tornes 1988). Brooding by female redwings is common soon after hatching but is by no means nearly constant (chapter 2), and at any rate, male redwings neither brood nor provision at all during this period (Yasukawa et al. 1990, Patterson 1991). Thus, red-winged blackbirds have the characteristics of a species in which male parental care is not greatly important to female reproductive success, though there is empirical evidence that it does have some effect (chapter 2).

In certain polygynous species, predation may be quite important in lowering the cost of sharing male parental care (Temrin and Jakobsson 1988, Bensch and Hasselquist 1991). The effect occurs because polygynous males typically help the most advanced brood on their territory almost exclusively, and therefore a secondary female is likely to fall heir to the male's assistance if (and only if) the first female's nest is depredated. As a result, the cost of polygyny decreases as the probability of nest predation increases. In red-winged blackbirds, rates of nest predation are high (chapter 4), and males have been observed to switch their provisioning effort to a later brood when an earlier one is depredated (Yasukawa et al. 1993), so predation undoubtedly does lower the cost of polygyny to some extent. The effect may not be as great as in some other species, however, in part because male provisioning is not that important a resource in redwings, and in part because male redwings are fairly likely to apportion some of their help to later broods even if earlier ones are not depredated (chapter 2).

Another factor that can act to lower the cost of sharing male parental care is asynchronous settlement of females within territories (Verner 1964, Leonard 1990). Asynchronous settlement has this effect because the less overlap there is between the nestling periods of females breeding on a territory, the better able the territory owner is to aid each of them. This factor also does not seem highly important in red-winged blackbirds, again in part because the cost of sharing male provisioning is not high, and also because settlement by females within territories is in fact often fairly synchronous (Orians 1980, Yasukawa and Searcy 1981, Westneat 1992b).

Other sources of a cost of polygyny, besides sharing of male parental care, also seem likely to be low in red-winged blackbirds. Proposals for such costs include the following. (1) Increased competition for food on the territory. Female redwings often forage off the territory, which should reduce any effects of competition for food. Indeed, extraterritorial foraging may be an adaptation to reduce such competition. (2) Interference from other females. Female red-winged blackbirds do interact aggressively within harems, but this aggression does not seem to restrict females to particular parts of a male's territory (chapter 9). There is no evidence that female redwings interfere with each other's eggs and young, even though we have watched unguarded nests in attempts to reveal such interference (Searcy unpublished). There is also no evidence of intraspecific brood parasitism ("egg dumping") (Gibbs et al. 1990, Harms et al. 1991, Westneat 1993a). (3) Reduced fertility of the male. The incidence of infertility, whether as the fault of the male or female, is very low in red-winged blackbirds, and there is no evidence that infertility increases with increasing harem size.

Although there are reasons to think that all proposed sources of a cost of polygyny are low in redwings, there is also solid evidence for a benefit of polygyny, stemming from a reduced risk of nest predation. This benefit has been demonstrated in experiments with artificial nests, which have shown that artificial clutches are safer from predation when placed near nests of female red-winged blackbirds (Ritschel 1985, Picman et al. 1988) and that artificial nests can make the natural nests safer as well (Ritschel 1985). Further evidence for an antipredation benefit of polygyny comes from female-removal experiments, which have shown that nest predation tends to increase when harem sizes are lowered (chapter 6), and from analyses showing that synchrony in nesting tends to lower predation (Westneat 1992b). One reason that this antipredation benefit is important in red-winged blackbirds is that nest predation is so important to female reproductive success (chapter 4). Another reason may be that male territories are often quite small, so that females nesting on the same territory are close enough to each other to make communal nest defense and predator dilution effective.

THE COST OF POLYGYNY IN OTHER SPECIES OF BIRDS

Here we ask, do the factors that act to lower the costs and increase the benefits of polygyny in red-winged blackbirds also apply to other polygynous species? Answering this question will lead us to consider how general is our no-cost explanation of polygyny in red-winged blackbirds.

Of the factors discussed above that act to lower the cost of polygyny in red-winged blackbirds, the most important must be the low level of non-shareable parental care provided by male red-winged blackbirds. Among other polygynous species, males in some resemble redwings in making relatively little contribution to provisioning of young, whereas in others males make much larger contributions, on the order of what is provided by males in monogamous species. The available data are summarized in Table 10.2. In bobolinks, pied flycatchers, the Polish population of great reed warblers, and house wrens, males perform close to 50% of the feeding visits at nests of monogamous or primary females. Contrasting species, in which males provide less that 25% of the provisioning at nests of primary females, are yellow-headed blackbirds and fan-tailed warblers in addition to redwings. A few populations are intermediate in terms of male contribution to parental care, notably savannah sparrows and the Japanese population of great reed warblers.

We have also given data in Table 10.2 on how the contribution of males to provisioning changes with female status. The difference in the percentage provisioning contributed by males at nests of secondary females versus primary and monogamous females together gives an idea of how much help a female is sacrificing when she mates with an already-mated male. Not all the studies report data in the same way, which complicates comparisons. It is clear, however, that the sacrifice made by secondary females is high in those populations, such as house wrens, pied flycatchers, and European great reed warblers, in which males contribute substantial help to primary females. Conversely, in populations such as fan-tailed warblers and western red-winged blackbirds, in which male contribution to primary females is very low or nonexistent, the sacrifice made by secondary females is also minimal.

For ten populations of eight species, we have data on both the contribution of males to feeding offspring (Table 10.2) and the degree of polygyny (Table 1.2). The relationship between these two variables is illustrated in Figure 10.1. Degree of polygyny is measured in this instance as the percentage of mated males that mate polygynously. Male contribution to feeding offspring is measured as the percentage of the total feeding visits made by the male at nests of monogamous or monogamous and primary females. Some of the data may not be independent, for example because one species contributes two points, or because some species are members of a single clade, so we do not give a test of statistical significance. Nevertheless, there is clearly a negative correlation between these two variables. The degree of polygyny decreases as male contribution to provisioning increases.

In particular, it appears that it is only in those species in which the contribution of males to provisioning is low (<25%) that polygynous matings predominate (>50%). Besides red-winged blackbirds, the two species showing this combination of characteristics are fan-tailed warblers and yellow-headed blackbirds; these, then, are the two species to which our no-cost

Table 10.2

Male Contribution to Feeding Nestlings in Various Species of Polygynous Passerines at Nests of Mongamous, Primary, Secondary, or Tertiary and Later Females.

Species	Locale	Male Contribution to Feeding Visits[a]				Reference
		Monogamous	Primary	Secondary	Tertiary+	
Yellow-headed blackbird	Washington	—	24	5	—	Willson 1966
	Manitoba	—	12	4	1	Lightbody & Weatherhead 1987a
Bobolink	Oregon	—	43	—	—	Wittenberger 1982
Fan-tailed warbler	Japan	0	0	0	—	Ueda 1984
Great reed warbler	Poland	53	41	4	—	Dyrcz 1986
	Japan	31	17	7	—	Urano 1990
Pied flycatcher	Sweden	51	53	29	—	Alatalo et al. 1982
	Norway	47	42	24	6	Lifjeld & Slagsvold 1989
Blackpoll warbler	New Brunswick	53	—	16	—	Eliason 1986
Savannah sparrow	New Brunswick	31	—	—	—	Wheelwright et al. 1992
House wren	Wyoming	65	63	28	—	L. Scott Johnson pers. comm.
Red-winged blackbird	Wisconsin	—	17	7	0	Yasukawa et al. 1990
	Washington	—	~1	<1	—	Beletsky & Orians 1991

[a]Contribution is measured as the percentage of feeding visits made by males. If nests of monogamous females are not listed separately, they are lumped with nests of primary females.

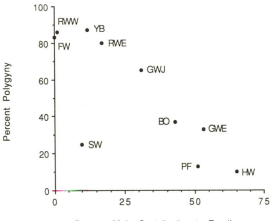

Figure 10.1 Relationship between percent polygyny and percent male contribution to feeding for various species of passerines. Percent polygyny is the percentage of territorial males that mate polygynously. Percent male contribution to feeding is the percentage of feeding trips to nests of monogamous or primary females made by males. An inverse relationship between degree of polygyny and male contribution to feeding young is clearly evident. FW = fan-tailed warbler (Ueda 1984), RWW = red-winged blackbird western population (Orians and Beletsky 1989, Beletsky and Orians 1990), SW = sedge wren (Burns 1982), YB = yellow-headed blackbird (Lightbody and Weatherhead 1987a), RWE = red-winged blackbird eastern population (Yasukawa et al. 1990), GWJ = great reed warbler Japanese population (Urano 1990), BO = bobolink (Wittenberger 1982), PF = pied flycatcher (Alatalo et al. 1982), GWE = great reed warbler European population (Dyrcz 1986, Catchpole 1986), HW = house wren (Johnson and Kermott 1991, L. Scott Johnson pers. comm.).

explanation of polygyny is most likely to generalize. Indeed, Lightbody and Weatherhead (1987a, 1988) have previously argued in favor of a no-cost model as the explanation for polygyny in one of these species, the yellow-headed blackbird. Lightbody and Weatherhead's (1988) "neutral-mate-choice" hypothesis assumes no cost and no benefit of polygyny, and random choice of breeding situations. As evidence against a cost of polygyny in their Manitoba study population, they show (1) that male assistance is low and has no effect on the number of young raised to fledging (Lightbody and Weatherhead 1987a), (2) that females mating with already-mated males are as successful in raising young as are monogamous females (Lightbody and Weatherhead 1987a), and (3) that the timing and spacing of settlement of females are not influenced by the presence of already-resident females (Lightbody and Weatherhead 1987b). In these respects, the mating system of yellow-headed blackbirds in this Manitoba population is very similar to the picture we have drawn for red-winged blackbirds, though there have been no direct tests of an antipredator benefit of nesting in harems. It appears, however, that these female yellow-headed blackbirds are much less influenced by environmental features in choosing where to settle than are redwing females (Lightbody and Weatherhead 1987a, 1988). Interestingly, there may be more of a cost of polygyny in Washington populations, where male yellow-headed blackbirds make a greater contribution to feeding off-spring (Willson 1966, see Table 10.2), and male help does affect the number of young raised and their mass at fledging (Patterson et al. 1980).

In the fan-tailed warbler, Ueda (1984) found that the only contribution males make to parental care is to build the outer fabric of the nest; this task is completed before the female even settles. Males perform no incubation, no brooding, no feeding of the female, and no provisioning of fledglings. Furthermore, in 98 hours of observation at 45 nests, Ueda (1984) never observed males to bring food to nestlings. Females nesting at the end of the breeding season are able to raise young after their mates have abandoned their territories (Ueda 1984). This certainly seems to be a no-cost system, though there are as yet few data available on female reproductive success. It is not known whether there is any benefit to polygyny, nor whether females show directed choice.

Other candidates for a no-cost explanation come from those species with intermediate levels of male provisioning, notably savannah sparrows and Japanese great reed warblers. In both these populations, males provide 31% of the feeding visits at monogamous nests (Wheelwright et al. 1992, Urano 1990). In savannah sparrows, Wheelwright et al. (1992) reported that secondary females received less male help than primary females, but not significantly less. Secondary females were able to compensate by working harder, so that broods without male assistance received as many total feeding visits as broods with male assistance. Male feeding rates had no effect on survival of nestlings nor on their mass at fledging. In one year out of four, young raised in polygynous associations were less likely to return than were young raised by monogamous pairs. If there is a cost of polygyny in savannah sparrows, it seems to be restricted to occasional years.

Urano (1990) found that in Japanese great reed warblers, males aided secondary females less than they did primary and monogamous females, though the differences were relatively small (Table 10.2). Nestlings of secondary females were no more likely to starve than were nestlings of monogamous and primary females, and were just as heavy at fledging. Secondary females compensated to some degree for reduced male assistance by working harder themselves, but this had no measurable effect on their survival. These findings contrast with the situation found by Dyrcz (1986) in a European population of this species, in which there was a much larger difference in male help received by primary and secondary females (Table 10.2), and young of secondary females were more likely to starve, and fledged at lower mass, than young of primary and monogamous females. Urano (1990) attributed the reduced cost of polygyny in Japan to (1) higher ambient temperatures, which reduced the need for brooding nestlings, freeing secondary females for provisioning, and (2) increased asynchrony in nesting attempts between primary and secondary nests, making it easier for males to provision both. In Japan, the average interval between nest starts of primary and secondary females on the same territory was 14 days (Urano 1990), whereas in Europe the interval in most cases was less than four days (Dyrcz 1986).

In contrast, there is good evidence of a cost of polygyny for certain of the species in which males make a large contribution to feeding the young. One example is the pied flycatcher. Antipredator benefits of polygyny are unlikely in this polyterritorial species, as the females mated to one male reside on separate territories, making both mutual defense and predator dilution unlikely. Secondary females receive substantially less help from their mates with feeding offspring than do monogamous and primary females (Table 10.2) and are unable to compensate fully by making more deliveries themselves (Alatalo et al. 1982). Secondary females are more likely to lose young to starvation (Alatalo et al. 1981) and only manage to raise 60 to 85% as many fledglings as simultaneously-nesting monogamous females (Alatalo and Lundberg 1984, Stenmark et al. 1988). Similarly, in house wrens, secondary females do not fully compensate for lost male assistance (Johnson and Kermott 1993) and produce fewer and lighter young than do primary and monogamous females (Johnson et al. 1993). Other species in which a cost of polygyny seems highly likely include bobolinks (Wittenberger 1982), blackpoll warblers (Eliason 1986), and northern harriers (Simmons et al. 1986).

The one species that is an obvious outlier in the graph relating degree of polygyny to the amount of male parental care is the sedge wren (Figure 10.1), which shows much less polygyny than might be expected given the low contribution of males to provisioning young. The study cited, Burns (1982), determined mating status for the smallest sample of males (12) of any of the studies used in Figure 10.1, but Crawford (1977b) found an equally low incidence of polygyny in a larger sample of sedge wrens. We have not included the sedge wren's congener, the marsh wren, in Figure 10.1 because no results have been published on the percentage contribution of males to feeding offspring in this species. Nevertheless, we suspect that marsh wrens would fall close to sedge wrens in Figure 10.1, as marsh wrens have a similar degree of polygyny (Table 1.2), and the best quantitative study of male feeding found only nine of 65 males (14%) feeding young (Leonard and Picman 1988), which implies that the average contribution of males must be quite low. Along with low male parental effort, marsh wrens also show pronounced asynchrony in the nesting cycles of females resident on the same territory (Verner 1964, Leonard and Picman 1987), which should further reduce any cost of sharing the male's parental effort (Leonard

1990). Thus, the curious thing about marsh wrens and sedge wrens is not that they are polygynous but rather that they show as low a degree of polygyny as they do. There must be other constraints on the degree of polygyny in these species, perhaps from female-female aggression or intraspecific nest destruction (Picman 1977b).

In conclusion, no-cost explanations of polygyny seem quite likely to apply to other species that resemble red-winged blackbirds in having high degrees of polygyny and low male parental effort, such as yellow-headed blackbirds and fan-tailed warblers. No-cost explanations may also apply to species or populations with intermediate degrees of polygyny and male parental effort, such as savannah sparrows and Japanese great reed warblers, and to species with low male parental care and low degrees of polygyny, such as sedge wrens and marsh wrens. No-cost explanations are unlikely to apply to species in which males make large contributions to nonshareable parental care; this category probably includes the majority of polygynous species, but mainly ones in which the degree of polygyny is low.

LONG-TERM MODELS

As we have stressed before, explanations of polygyny such as the no-cost model seek to explain why females enter into polygynous associations now, under present-day conditions. In these models, vital aspects of present-day conditions, such as the average male contribution to parental care or the variance in territory quality, are regarded as fixed, and only female choice is allowed to change. Thus, these hypotheses address the maintenance of polygyny rather than its origin.

In chapter 6 we briefly considered long-term models, which do address the evolutionary origins of polygyny, and which do allow evolution of important conditions, such as average male parental care. In that chapter we were not successful in settling on a single long-term model to explain the origin of polygyny in red-winged blackbirds, so we are not in a good position to generalize a redwing answer to other species. The above discussion of the cost of polygyny in various species, however, raises one possibility that we would like to consider here: that a no-cost hypothesis can be extended to explain the origin as well as the maintenance of polygyny.

We have shown that the degree of polygyny across species is closely related to male contribution to provisioning young, and in particular that a high degree of polygyny only occurs in species in which male contribution is low. This relationship argues that, in highly polygynous species, polygyny now has no cost or at most a low cost. This need not mean, however, that when polygyny originated in these species, there was then no cost to polygyny. We have argued previously that the cost of polygyny and the degree of polygyny can coevolve: as the degree of polygyny increases, males are selected to put more time and energy into mate attraction, mate guarding, and territoriality, and less into parental care; lower male parental care lowers the cost of polygyny, selecting for more polygynous mating among females, etc. (see Figures 6.1 and 6.2). It seems likely to us, then, that highly polygynous species such as red-winged blackbirds or fan-tailed warblers were at one time weakly polygynous, and that at that time polygyny had some cost, as it seems to have in most species that are now weakly polygynous. These species then for some reason entered the coevolutionary spiral in which the degree of polygyny increases and the cost of polygyny decreases. Therefore, we think that no-cost models are unlikely to apply to the origin of polygyny, but instead apply only to its maintenance.

Sexual Selection and Other Effects of Polygyny

Male red-winged blackbirds vary greatly in mating success, due in large part to the species' polygynous mating system (chapter 7). One component of the variance in mating success is a difference in success between territory owners and nonowners. This component can be ascribed to polygyny, because a system in which some males obtain many mates inevitably produces a surplus of males with no mates, as long as the sex ratio is roughly equal. A second component is the variance in mating success among territorial males caused by differences in harem sizes; this component is obviously inseparable from social polygyny. The remainder of the variance in male mating success derives from differences in success in achieving copulations, both intra- and extrapair; this component cannot be ascribed to social polygyny, as the same kinds of differences can occur in monogamous species.

Our first major conclusion about sexual selection is that current sexual selection in male red-winged blackbirds is weak (chapter 7). Supporting this conclusion are a number of studies comparing males that succeed in obtaining and defending a territory with those that fail. These studies have produced results that are almost entirely negative; consistent differences between successful and unsuccessful males cannot be demonstrated. The conclusion is also supported by studies relating traits of territory owners to their harem sizes. Again, a number of such studies have been done and have not succeeded in finding consistent relationships between male traits and harem sizes. The major gap in the evidence for weak sexual selection is that no results relating male traits to success in copulation have yet been published; therefore our conclusion that sexual selection is weak actually applies to sexual selection exclusive of this last component.

Weak current sexual selection is not a general result; much stronger sexual selection in progress has been found in a number of other species. Examples include species of fish (Downhower and Brown 1980, Hastings 1988), amphibians (Howard 1980, 1981, Arak 1988), and lizards (Ruby 1981) in which there is current sexual selection favoring large size; species of fish (Kodric-Brown 1983) and birds (Møller 1988, Hill 1990) in which there is sexual selection for visual display traits; and species of insects (Snedden and Sakaluk 1992) and birds (Gibson and Bradbury 1985, Catchpole 1986, Radesäter et al. 1987) in which there is sexual selection for vocal display traits. It is interesting to consider why sexual selection in progress is stronger in these cases than in red-winged blackbirds.

We have already suggested two factors that weaken current sexual selection in redwings: the lottery aspects of territory ownership and self-limiting sexual selection. The lottery aspects of territoriality stem in large part from resident's advantage, the advantage an owner has in contesting for his territory just by virtue of being the owner. As a result of resident's advantage, nonowners may not be able to take a territory away from an inferior owner, and there may be little of the shuffling of territories among owners that is needed if each is to end up with a territory commensurate with his own quality. Resident's advantage may explain the strong site fidelity that has been demonstrated for territorial male redwings.

Resident's advantage itself appears to be quite a common phenomenon among territorial species, occurring even in species with very ephemeral territories (Davies 1978). Of course, resident's advantage cannot operate to weaken sexual selection in species without territories. This may explain why sexual selection is stronger in species such as wood frogs (Howard 1980) and yellow-rumped caciques (Robinson 1986) in which males struggle directly for control of females. Resident's advantage should also not weaken sexual selection in species which are territorial but in which females choose males directly on the males' own traits, rather than on territory traits, as may be the case in sage grouse (Gibson and Bradbury 1985) and other lekking species, and in some passerines where females seem to choose on song features (Catchpole 1980, 1986, Radesäter et al. 1987). In redwings, males compete with one another for control of territories, and females base settlement choices largely on territory quality; it is only under these circumstances that lottery competition can weaken sexual selection.

The second factor that we have suggested weakens sexual selection is self-limiting sexual selection. Sexual selection is usually pictured as working in a directional manner, favoring the extreme of some trait. Since we do not expect most traits to be continually evolving more extreme values, we also usually picture the trait as being held in balance by some opposing selective force, usually survival selection. It is under this kind of scenario that we expect to find a trait that has evolved due to sexual selection to be still subject to strong sexual selection in progress. It is possible, however, for sexual selection to be self-limiting, favoring some stable value of a trait rather than an ever more extreme value. Under this latter scenario, traits that have evolved due to sexual selection would no longer be subject to directional selection.

We have proposed that self-limiting sexual selection occurs specifically with regard to size in male red-winged blackbirds. The idea is that large size is advantageous in direct physical contests during territory defense, but that small size is advantageous in aspects of territory defense requiring endurance, that is, the ability to sustain energy-costly behaviors over long periods. Because both large and small individuals have advantages in competition for territory, there is no directional selection favoring either extreme in size. Self-limiting sexual selection may be most likely in such cases, in which both fighting ability and endurance contribute to success; however, it is also necessary that energy demands are high enough and last long enough that they cannot be met using stored reserves alone, as otherwise large males would have an advantage in endurance as well as in fighting.

A third factor weakening sexual selection in progress is the limited natural variability that exists in some of the traits associated with sexual selection in male redwings. Sexual selection in progress is often demonstrated by a correlation between a male trait and mating success. Such a correlation expresses the amount of variation in mating success that can be explained by variance in the male trait. Other things being equal, we would expect that the more variable is the male trait, the more variance in mating success can be explained by that trait. An analogy may make this point clearer. In sports such as boxing and wrestling, contestants compete within weight classes, which results in relatively little variance in weight among competitors. Under these circumstances, weight presumably would not be much help in explaining or predicting success. If instead we had contestants of all weights competing together, weight would then be an important predictor of success.

Size is one example of a trait with limited natural variation among adult male red-winged blackbirds. In a sample of 42 male redwings holding territory in one year in a Washington population, Searcy (1979b, unpublished) found wing lengths varying from 125 to 136 mm; thus the range in wing lengths was 8% of the maximum. In two other years the range was 7 and 9% of the maximum. Correlations between wing lengths and harem sizes were consistently low. By contrast, Howard (1980) found a strong correlation between size (snout-ischium length) and the probability of mating in a population of wood frogs, and here the range in size was 34 to 46 mm, or 26% of the maximum. In bullfrogs, male size was again strongly correlated with mating success within a population, and the range in size among competing males was 25 to 40% of the maximum, depending on the year (Howard 1981). In the angel blenny, Hastings (1988) found that male reproductive success was significantly correlated with body length in a population in which the range in length among mature males was 55% of the maximum. In yellow-rumped caciques, Robinson (1986) found that heavier males tended to be dominant and that dominant males had higher mating success; here the range in weights among competing males was approximately 27% of the maximum. Note that many of the cases in which there is high variance in male size occur in animals, such as fish, amphibians, and reptiles, that are indeterminate growers.

Redwings also show comparatively low variability in visual display traits, though this is harder to quantify. The color of the epaulets of adult male red-winged blackbirds varies from red-orange to red, and color does not correlate consistently with harem size (Searcy 1979b). By contrast, Hill (1990) found that male house finches vary in color over a much larger range, from pale yellow to intense red, and here color is significantly related to pairing success. Red in the male redwing is confined to two discrete patches of relatively standardized size; the range in length of red is approximately 20 to 30% of the maximum (Searcy unpublished). This is greater variability than shown by redwings in body size but is much lower than the variability in extent of coloration in some species. For example, Houde (1987) found male guppies in one population had orange pigment covering 0 to 30% of their bodies, making the range in extent of coloration 100% of the maximum. Natural variation in the extent of color correlates with mating success in guppies (Houde 1987) but not in redwings (Searcy 1979b, Shutler and Weatherhead 1991a). In at least some of the cases in which color patterns are highly variable, coloration is to a large extent condition dependent, and so may be more environmentally than genetically determined (Hill 1992).

Our argument that male traits associated with sexual selection show low variability in redwings does not extend to vocal display traits. Yasukawa et al. (1980) found repertoires of from two to eight song types within a population, giving a range that is 75% of the maximum. Repertoire size does correlate with pairing success in redwings, though the correlation is much weakened when other factors, such as male age, are controlled. Variation in aggressiveness is difficult to quantify.

If we are right about the factors weakening sexual selection in progress in red-winged blackbirds, then we should expect to find weak sexual selection in progress in species that show: (1) low variance in male traits associated with fighting and display; (2) a mating system characterized by male territoriality, female choice based on territory quality, and resident's advantage; and/or (3) competition among males that involves both fighting and endurance. The first characteristic, low variance in male traits, is especially likely to be found in species that are determinate growers, such as most birds and mammals, and in which display traits are not strongly condition dependent. A mating system in which females choose on male territories is most likely to be found in species in which females breed on the male's territory, so that features of the territory are important to their reproductive success. Competition among males involving both fighting and endurance seems most likely in species in which males hold territories for long periods.

We expect, then, that the combination of weak sexual selection in progress with considerable variation in male mating success will be found in determinate growers lacking condition-dependent display traits and showing a mating system with prolonged defense by males of territories used by females for breeding. This prediction may seem excessively limited, but in fact it is appropriately applied to a great many bird species, as well as at least some mammals.

Our second major conclusion regarding sexual selection is that male red-winged blackbirds posses a number of traits that are adaptations evolved due to sexual selection. One such trait is the epaulet (chapter 8). Evidence that the epaulet functions in enhancing mating success includes the observation that epaulets are prominently displayed in aggressive postures associated with defense of territory and in courtship (Orians and Christman 1968). More critical is experimental evidence showing that territory owners whose epaulets are blackened suffer a greatly increased chance of losing their territories (Peek 1972, Smith 1972), and other experiments showing that the success of taxidermic mounts in keeping territories free of intruders increases with the size of the mounts' epaulets (Røskaft and Rohwer 1987).

A second trait that appears to be an adaptation evolved through sexual selection is male song. This conclusion is supported by the observation that song is used by males both when courting females and during aggressive encounters in the defense of territory. The importance of song in territory defense has been shown by experiments in which surgically muted territory owners were found to suffer an increased probability of territory loss (Peek 1972, Smith 1979) and by other experiments in which song broadcast from loudspeakers was found to discourage rival males from entering the territory (Yasukawa 1981b, c). The importance of song in courtship has been demonstrated by experiments showing that playback of song stimulates females to perform the courtship posture that precedes copulation (Searcy 1988b).

There is also evidence that particular aspects of singing behavior have evolved due to sexual selection. The best case can be made for the song repertoire. Experiments have shown that playback of multiple song types is superior to playback of single song types in limiting certain types of intrusion onto territories (Yasukawa 1981b). Other experiments have shown the multiple song types are also superior in eliciting courtship from females (Searcy 1988b). Finally, there is experimental evidence that males greatly increase their use of the repertoire when courting females, i.e., males cycle through their song types more rapidly during courtship than they do when singing undisturbed (Searcy and Yasukawa 1990).

Larger male than female size may be another adaptation evolved due to sexual selection, but here the evidence is less compelling. Ideally, a test of sexual selection for size dimorphism would employ an experimental manipulation of male and female size. Unfortunately, no one has succeeded in manipulating size, the way that epaulets and song have been manipulated, so that experimental tests of the importance of large male size in territory defense and courtship have not been made. Instead, the primary evidence that size dimorphism has evolved due to sexual selection comes from comparative data, showing that size dimorphism is positively associated with harem size among icterine species (Webster 1992). The comparative evidence is as convincing as it could be for this association, in that there is a strong relationship that holds even if phylogeny and body size are controlled. We feel that by its nature, however, comparative evidence can never be as conclusive as the kinds of experimental evidence available for visual and vocal display.

Aggressiveness in males may be another sexually-selected trait, but here the evidence is again inconclusive. Patterns of use support an intrasexual function of aggressiveness, in that both aggressive display and fighting occur mainly in the context of territory defense. Experimental manipulations of aggressiveness using hormones and hormone blockers, however, have given somewhat ambiguous results, perhaps because the treatments are not that successful in altering aggressive behavior.

The conclusion that vocal and visual display traits are adaptations for sexual selection is undoubtedly very general. Visual display traits, such as patches of coloration, influence mating success in species from fish (Houde 1987) to lizards (Thompson and Moore 1991), and from butterflies (Krebs and West 1988) to birds (Hill 1990). Male songs apparently function in enhancing mating success not only in birds but in anurans and acoustic insects as well (Searcy and Andersson 1986). Song repertoires function in stimulating female courtship in a variety of species of birds (Searcy 1992). Nevertheless, there are some interesting exceptions, where display traits do not have the expected functions. As one example, Rohwer and Røskaft (1989) found that blackening the heads and breasts of male yellow-headed blackbirds did not lower the ability of these males to hold territories and attract mates; in fact, substantially more of the blackened males traded up for better territories than among the controls. This species is another icterine, and thus a near relative of red-winged blackbirds, and its yellow head seems closely analogous to the redwing's red epaulet as a visual display trait (although the yellow head is not coverable). The fact that the functions of these traits must be entirely different makes the point that we need to be cautious in generalizing functional interpretations across species.

Note that two of our explanations for weak sexual selection in progress are easy to reconcile with the existence of sexually-selected adaptations, whereas the third is not. Self-limiting sexual selection is obviously compatible with the occurrence of sexually-selected adaptations, as this hypothesis assumes that sexual selection can be strongly directional over some range of trait values and stabilizing or weakly directional at more extreme values. Thus, sexual selection could have acted directionally while the adaptation evolved, while now acting to stabilize the trait. The idea that low variance in traits subject to sexual selection leads to weak sexual selection can also be reconciled with the existence of sexually-selected adaptations. For example, it is possible that a trait such as body size originally had a larger variance as well as a lower mean, and that directional sexual selection caused the variance to decrease as it caused the mean to increase. This scenario is just another example of Fisher's (1930) view of the Fundamental Theorem of Natural Selection, in which the response to selection occurs at the expense of heritable variance.

More difficult to reconcile with the existence of sexually-selected traits is our argument about lottery competition for territories. There is no reason to think that the territorial system would change as traits such as epaulets, song, or size changed, so if the territorial system weakens sexual selection now, it presumably also weakened sexual selection during the evolution of these traits. If so, and if we are right about lottery competition, then traits such as epaulets and song would have to evolve due to relatively weak sexual selection.

The major effect of polygyny on female red-winged blackbirds is not that it increases the opportunity for sexual selection but that it increases the opportunity for social interaction among females. Female redwings do vary in the number of mates during a breeding season, but this variation must have at most a minor impact on female reproductive success, which is limited by the number of young for which a female is able to care, rather than by the number of matings she obtains. Moreover, if female choice of males with which to copulate causes differential reproductive success among females, this component of sexual selection cannot be ascribed to polygyny because such female choice also occurs in some monogamous species.

Polygyny definitely does raise the frequency of interaction among females during the breeding season, and female red-winged blackbirds have consequently evolved adaptations that function in female-female interactions. We do not mean to imply that the female social structure within harems is itself such an adaptation. Harems do show a fairly elaborate social structure, with dominance relationships between females that depend in part on their order of settlement, and which change according to where an interaction takes place relative to the nests of the interactants (chapter 9). For the most part, however, this elaborate social structure may be the outcome of rather simple individual behaviors, in which individuals compete for access to resources with levels of confidence that vary according to their length of residency and location relative to their home base. We see little reason to think that these aggressive/dominance behaviors have changed due to polygyny; females in monogamous species might behave similarly if put in similar situations.

One characteristic of female red-winged blackbirds that arguably is an adaptation for polygyny is their complex vocal repertoire (chapter 9). Whereas females in many passerine species do not sing at all, female red-winged blackbirds have a variety of song types, and they sing very frequently during the breeding season. One of their song types, the teer, is given mainly to other females, and in general its pattern of use suggests an aggressive function. A playback experiment with this type of song, however, suggests its effect may be to attract settlement by other females (Yasukawa 1990). The teer certainly seems to be an adaptation for female-female interaction, but more work is needed to determine its exact function. A second song type, the chit, is given by females at their nests and in reply to males, and it seems to function in deterring harassment from the male and in directing his attention to a female's nesting attempt. The chit can be considered an adaptation to polygyny, in that its functions in reinforcing male-female relations are made necessary by the fact that a female has to compete with other harem members for the territory owner's attentions.

It is tempting to consider the epaulets and chin coloration of female red-winged blackbirds as adaptations for signaling aggression and dominance in the context of the heightened social competition within harems; however, there is no evidence that these female visual displays do function in this way. There is some evidence that social competition within harems selects for larger size in females, thus acting to lessen sexual size dimorphism. Female aggressiveness may also be in part an adaptation to increased social competition, but so far evidence on the function of female aggression is contradictory. We conclude that the clearest adaptation of female red-winged blackbirds to social polygyny is their complex repertoire of songs.

Building a Better Redwing: Prospects for Future Research

One way to identify avenues for future research is to find complete gaps in our knowledge. One such gap for red-winged blackbirds, which we have mentioned often before, concerns the relationship between male traits and female choice in copulation. This gap can be filled by studies in which male traits are measured, copulations are observed, and paternity of the young is determined using DNA methods. It will be interesting to learn from such studies whether this component of current sexual selection is stronger than the others that have been measured thus far, and if so, what male traits are favored. Other gaps in our knowledge concern the behavior of floater males and the tactics they use to obtain a territory, and the searching behavior of females choosing a territory for the first time. These gaps exist because these classes of individuals are difficult to catch and mark, and even more difficult to follow and observe. Some progress has been made using radiotracking of floaters in redwings (Dave Shutler and Patrick J. Weatherhead pers. comm.) and of prospecting females in other species (Bensch and Hasselquist 1992); this seems the most promising technique. Following floaters would help test the lottery aspects of territoriality, for example by determining whether the first male to reach a vacancy does have an advantage. Following prospecting females would allow analysis of the decisions rules used by females in mate choice, rules such as a best-of-n strategy (visit n males and then return to the best one) and sequential sampling (establish a threshold criterion and take the first male encountered that exceeds it; see Janetos 1980, Real 1990). This is an aspect of sexual-selection theory that has been barely touched on so far in work on redwings.

A second way to identify areas in need of research is to find contradictory conclusions in previous studies. An example for redwings concerns the effects of resident females on further female settlement. We believe that the bulk of the evidence for redwings indicates residents have no deterrent effect on settlement; especially convincing to us are the various female removal experiments of Searcy (1988a, unpublished) and the speaker-occupation experiment of Yasukawa (1990). One carefully designed and conducted removal experiment (Hurly and Robertson 1984), however, did support a deterrent effect of residents on settlement. All the studies cited here used different methodologies, so it may be that the discrepancy is just an artifact. A more interesting possibility is that the discrepancy reveals a real difference in behavior, due perhaps to large-scale geographic differences in behavior and ecology, or to more local peculiarities of study sites. For example, it may be that residents deter settlement where there is greater competition for food on the territory, or where mutual nest defense and satiation of nest predators are less important. These and other possibilities would reward further investigation.

A third way of identifying areas in need of research is to find aspects of our picture of redwing behavior and ecology that seem paradoxical. In general, behavioral ecologists try to understand how a species' behavior functions to promote individual survival and reproduction. A paradox arises when behavior appears to be maladaptive, or at least suboptimal. This is not to say that we believe all behavior has to be adaptive and optimal; in fact, we have argued that certain aspects of redwing behavior, such as female response to song repertoires, may be nonadaptive. It does seem to us paradoxical, however, when a species shows a behavior that does not seem adaptive, and where a simple change in behavior would arguably improve adaptation. By a simple change in behavior we mean one that ought to be within the species' abilities. We can identify such paradoxes by asking what changes we could make to build a better redwing, hence the title of this section.

One change that might result in a better redwing would be to make females less aggressive. We have seen that female red-winged blackbirds perform aggressive displays and attack each other during the breeding season. Aggression is shown against fellow harem members but is directed especially at outsiders showing interest in settling on the territory. Much of the aggression seems aimed at discouraging settlement by these prospecting females. The paradox is that there is little evidence that female aggression actually does discourage settlement, or that residents would benefit if they did succeed in discouraging settlement. Females may actually benefit from each other's presence: why, then, aren't they more tolerant of each other? Tolerance seems an easy trait to evolve; it merely requires showing less aggression. Clearly, there is something here that we do not understand.

Another change that would seem likely to make a better redwing is to make males more aggressive, as Shutler and Weatherhead (1992) have argued. This applies especially to floater males. As we have seen, a large proportion of the male population "floats" without a territory. Many of the floaters are adults that do not seem in any measurable way inferior to territory owners. Floaters seem to have zero reproductive success, and a high probability of dying before they can get a territory. The costs and benefits are such that we would expect selection to favor "desperado" tactics on the part of floaters, in which they would mount all-out aggressive attacks on owners in order to gain a territory (Grafen 1987, Shutler and Weatherhead 1992). Instead, what is actually observed is that any overt aggression of floaters toward owners is extremely rare. We can invent various rationales for the timidity of floaters: perhaps fighting is more dangerous than we think, or perhaps survival is higher among floaters than among territory owners, so that there is an advantage of remaining a floater that partially balances the cost. We do not find any of the rationales very satisfying, however. One intriguing possibility, which would resolve both our paradoxes at once, is that there is a genetic correlation between aggressiveness in males and females, as Muma and Weatherhead (1989) have suggested for plumage coloration. Genetic correlations between the sexes are expected to slow the evolution of sexual dimorphism (Lande 1980) and thus in this case would act to keep females too aggressive and males too passive. Testing for genetic correlations would be difficult but not impossible.

As we stated at the start of the book, one of the reasons that redwings have been studied so extensively is that there is a sort of snowball effect: when a little is found out about a species, more questions are revealed, more research is then done, more questions are raised, etc. This runaway process is by no means completed for red-winged blackbirds, and redwing researchers show no signs of self-limitation.

Common and Scientific Names

BIRDS

American crow, *Corvus brachyrhynchos*
black-billed magpie, *Pica pica*
blackpoll warbler, *Dendroica striata*
blue jay, *Cyanocitta cristata*
bobolink, *Dolichonyx oryzivorus*
Brewer's blackbird, *Euphagus cyanocephalus*
brown-headed cowbird, *Molothrus ater*
brown thrasher, *Toxostoma rufum*
common grackle, *Quiscalus quiscula*
corn bunting, *Emberiza calandra*
cliff swallow, *Hirundo pyrrhonota*
dickcissel *Spiza americana*
dunnock, *Prunella modularis*
fan-tailed warbler, *Cisticola juncidis*
gray catbird, *Dumetella carolinensis*
great reed warbler, *Acrocephalus arundinaceus*
great tit, *Parus major*
house finch, *Carpodacus mexicanus*
house martin, *Delichon urbica*
house sparrow, *Passer domesticus*
house wren, *Troglodytes aedon*
indigo bunting, *Passerina cyanea*
kingfisher, *Alcedo atthis*
marsh wren, *Cistothorus palustris*
meadow pipit, *Anthus pratensis*
mute swan, *Cygnus olor*
northern harrier, *Circus cyaneus*
northern mockingbird, *Mimus polyglottos*
osprey, *Pandion haliaetus*
pied flycatcher, *Ficedula hypoleuca*
prairie warbler, *Dendroica discolor*
red-billed gull, *Larus novaehollandiae*
red-tailed hawk, *Buteo jamaicensis*
red-winged blackbird, *Agelaius phoeniceus*
rock dove, *Columba livia*
sage grouse, *Centrocercus urophasianus*
savannah sparrow, *Passerculus sandwichensis*
Scott's seaside sparrow, *Ammodramus maritimus peninsulae*
sedge wren, *Cistothorus platensis*
shiny cowbird, *Molothrus bonariensis*
song sparrow, *Melospiza melodia*
splendid fairy wren, *Malarus splendens*
Tengmalm's owl *Aegolius funereus*
tree swallow, *Tachycineta bicolor*
tricolored blackbird, *Agelaius tricolor*
Virginia rail, *Rallus limicola*
western meadowlark, *Sturnella neglecta*
winter wren, *Troglodytes troglodytes*
wood warbler, *Phylloscopus sibilatrix*
yellow-headed blackbird, *Xanthocephalus xanthocephalus*
yellow-hooded blackbird, *Agelaius icterocephalus*
yellow-rumped cacique, *Cacicus cela*
yellow-shouldered blackbird, *Agelaius xanthomus*
yellow-winged blackbird, *Agelaius thilius*
zebra finch, *Taeniopygia guttata*

MAMMALS

beaver, *Castor canadensis*
Grant's gazelle, *Gazella granti*
harvest mouse, *Reithrodontomys megalotis*
long-tailed weasel, *Mustela frenata*
mink, *Mustela vison*
raccoon, *Procyon lotor*

red squirrel, *Tamiasciurus hudsonicus*
waterbuck, *Kobus ellipsiprymnus*

AMPHIBIANS

bullfrog, *Rana catesbiana*
frog, *Physalaemus pustulosus*
frog, *Physalaemus coloradorum*
wood frog, *Rana sylvatica*

FISH

angel blenny, *Coralliozetus angelica*
guppy, *Poecilia reticulata*

PLANTS

bulrush, *Scirpus* spp.
bunch grass, *Coix* spp. and *Paspalum* spp.
bur reed, *Sparaganium eurycarpum*
buttonbush, *Cephalanthus occidentalis*
cattail, *Typha* spp.
cord grass, *Spartina* spp.
mustard, *Brassica* spp.
purple loosestrife, *Lythrum salicaria*
sedge, *Cyperus* spp.
wild radish, *Raphanus sativus*
willow, *Salix* spp.

References

Alatalo, R. V., A. Carlson, and A. Lundberg. 1990. Polygyny and breeding success of pied flycatchers nesting in natural cavities. Pp. 323–330 in J. Blondel, A. Gosler, J. Leberto, and R. McCleary, eds., *Population Biology of Passerine Birds*. Springer-Verlag, Berlin.

Alatalo, R. V., A. Carlson, A. Lundberg, and S. Ulfstrand. 1981. The conflict between male polygamy and female monogamy: the case of the pied flycatcher *Ficedula hypoleuca*. *Am. Nat.* 117:738–753.

Alatalo, R. V., and A. Lundberg. 1984. Polyterritorial polygyny in the pied flycatcher *Ficedula hypoleuca*—evidence for the deception hypothesis. *Ann. Zool. Fenn.* 21:217–228.

Alatalo, R. V., and A. Lundberg. 1990. Polyterritorial polygyny in the pied flycatcher. *Adv. Study Behav.* 19:1–27.

Alatalo, R. V., A. Lundberg, and K. Ståhlbrandt. 1982. Why do pied flycatcher females mate with already-mated males? *Anim. Behav.* 30:585–593.

Alexander, R. D., J. L. Hoogland, R. D. Howard, K. M. Noonan, and P. W. Sherman. 1979. Sexual dimorphism and breeding systems in pinnipeds, ungulates, primates, and humans. Pp. 402–435 in N. A. Chagnon and W. D. Irons, eds., *Evolutionary Biology and Human Social Behavior: An Anthropological Perspective*. Duxbury Press, North Scituate, Mass.

Allen, A. A. 1914. The red-winged blackbird: a study in the ecology of a cat-tail marsh. *Proc. Linn. Soc. N. Y.* 24–25:43–128.

Altmann, S. A., S. S. Wagner, and S. Lenington. 1977. Two models for the evolution of polygyny. *Behav. Ecol. Sociobiol.* 2:397–410.

Andersson, M. 1994. *Sexual Selection*. Princeton University Press, Princeton, N. J.

Arak, A. 1988. Sexual dimorphism in body size: a model and a test. *Evolution* 42:820–825.

Arcese, P. 1987. Age, intrusion pressure and defence against floaters by territorial male song sparrows. *Anim. Behav.* 35:773–784.

Armstrong, E. A. 1973. *A Study of Bird Song*. Dover, New York.

Armstrong, T. A. 1992. Categorization of notes used by female red-winged blackbirds in composite vocalizations. *Condor* 94:210–223.

Arnold, S. J. 1983. Sexual selection: the interface of theory and empiricism. Pp. 67–107 in P. Bateson, ed., *Mate Choice*. Cambridge University Press, Cambridge.

Arnold, S. J., and M. J. Wade. 1984. On the measurement of natural and sexual selection: theory. *Evolution* 38:709–719.

Bacon, P. J., and P. Andersen-Harild. 1989. Mute swan. Pp. 363–386 in I. Newton, ed., *Lifetime Reproduction in Birds*. Academic Press, London.

Ball, R. M., Jr., S. Freeman, F. C. James, E. Bermingham, and J. C. Avise. 1988. Phylogeographic population structure of red-winged blackbirds assessed by mitochondrial DNA. *Proc. Natl. Acad. Sci. USA* 85:1558–1562.

Bart, J., and A. Tornes. 1989. Importance of monogamous male birds in determining reproductive success: evidence for house wrens and a review of male-removal studies. *Behav. Ecol. Sociobiol.* 24:109–116.

Bateman, A. J. 1948. Intrasexual selection in *Drosophila. Heredity* 2:349–368.

Beer, J. R., and D. Tibbitts. 1950. Nesting behavior of the red-winged blackbird. *Flicker* 22:61–77.

Beletsky, L. D. 1983a. Aggressive and pair-bond maintenance songs of female red-winged blackbirds. *Z. Tierpsychol.* 62:47–54.

Beletsky, L. D. 1983b. An investigation of individual recognition by voice in female red-winged blackbirds. *Anim. Behav.* 31:355–362.

Beletsky, L. D. 1985. Intersexual song answering in red-winged blackbirds. *Can. J. Zool.* 63:735–737.

Beletsky, L. D. 1990. Alert calls of male red-winged blackbirds: do females listen? *Behaviour* 23:1–12.

Beletsky, L. D. 1992. Social stability and territory acquisition in birds. *Behaviour* 123:290–313.

Beletsky, L. D., S. Chao, and D. G. Smith. 1980. An investigation of song-based species recognition in the red-winged blackbird *(Agelaius phoeniceus). Behaviour* 73:189–203.

Beletsky, L. D., and M. G. Corral. 1983. Lack of vocal mate recognition in female red-winged blackbirds. *J. Field Ornithol.* 54:200–202.

Beletsky, L. D., B. J. Higgins, and G. H. Orians. 1986. Communication by changing signals: call switching in red-winged blackbirds. *Behav. Ecol. Sociobiol.* 18:221–229.

Beletsky, L. D., and G. H. Orians. 1985. Nest-associated vocalizations of female red-winged blackbirds, *Agelaius phoeniceus. Z. Tierpsychol.* 69:329–339.

Beletsky, L. D., and G. H. Orians. 1987a. Territoriality among male red-winged blackbirds. I. Site fidelity and movement patterns. *Behav. Ecol. Sociobiol.* 20: 21–34.

Beletsky, L. D., and G. H. Orians. 1987b. Territoriality among male red-winged blackbirds. II. Removal experiments and site dominance. *Behav. Ecol. Sociobiol.* 20:339–349.

Beletksy, L. D., and G. H. Orians. 1989a. Territoriality among male red-winged blackbirds. III. Testing hypotheses of territorial dominance. *Behav. Ecol. Sociobiol.* 24:333–339.

Beletsky, L. D., and G. H. Orians. 1989b. Familiar neighbors enhance breeding success in birds. *Proc. Natl. Acad. Sci. USA* 86:7933–7936.

Beletksy, L. D., and G. H. Orians. 1989c. Red bands and red-winged blackbirds. *Condor* 91:993–995.

Beletsky, L. D., and G. H. Orians. 1990. Male parental care in a population of red-winged blackbirds, 1983–1988. *Can. J. Zool.* 68:606–609.

Beletsky, L. D., and G. H. Orians. 1991. Effects of breeding experience and familiarity on site fidelity in female red-winged blackbirds. *Ecology* 72:787–796.

Beletsky, L. D., G. H. Orians, and J. C. Wingfield. 1989. Relationships of steroid hormones and polygyny to territorial status, breeding experience, and reproductive success in male red-winged blackbirds. *Auk* 106:107–117.

Beletsky, L. D., G. H. Orians, and J. C. Wingfield. 1990. Effects of exogenous androgen and antiandrogen on territorial and nonterritorial red-winged blackbirds (Aves: Icterinae). *Ethology* 85:58–72.

Beletsky, L. D., G. H. Orians, and J. C. Wingfield. 1992. Year-to-year patterns of circulating levels of testosterone and corticosterone in relation to breeding density, experience, and reproductive success of the polygynous red-winged blackbird. *Horm. Behav.* 26:420–432.

Bell, G. 1989. A comparative method. *Am. Nat.* 133:553–571.

Bensch, S., and D. Hasselquist. 1991. Nest predation lowers the polygyny threshold: a new compensation model. *Am. Nat.* 138:1297–1306.

Bensch, S., and D. Hasselquist. 1992. Evidence for active female choice in a polygynous warbler. *Anim. Behav.* 44:301–311.

Bent, A. C. 1958. *Life Histories of North American Blackbirds, Orioles, Tanagers, and Allies.* U. S. Natl. Mus. Bull. 211. Dover, New York.

Berger, M., and J. S. Hart. 1972. Die Atmung beim Kolibri *Amazilia fimbriata* während des Schwirrfluges bei verschiedenen Umgebungstemperaturen. *J. Comp. Physiol.* 81:363–380.

Birkhead, T. R., and A. P. Møller. 1992. *Sperm competition in birds: evolutionary causes and consequences.* Academic Press, London.

Birks, S. M., and L. D. Beletsky. 1987. Vocalizations of female red-winged blackbirds inhibit sexual harassment. *Wilson Bull.* 99:706–707.

Björklund, M. 1990. A phylogenetic interpretation of sexual dimorphism in body size and ornament in relation to mating system in birds. *J. Evol. Biol.* 3:171–183.

Blakley, N. R. 1976. Successive polygyny in upland nesting redwinged blackbirds. *Condor* 78:129–133.

Blank, J. L., and V. Nolan, Jr. 1983. Offspring sex ratio in red-winged blackbirds is dependent on maternal age. *Proc. Natl. Acad. Sci. USA* 80:6141–6145.

Bray, O. E., J. J. Kennelley, and J. L Guarino. 1975. Fertility of eggs produced on territories of vasectomized red-winged blackbirds. *Wilson Bull.* 87:187–195.

Breitwisch, R., P. G. Merritt, and G. H. Whitesides. 1986. Parental investment by the northern mockingbird: male and female roles in feeding nestlings. *Auk* 103:152–159.

Brenner, F. J. 1966. The influence of drought on reproduction in a breeding population of redwinged blackbirds. *Am. Midl. Nat.* 76:201–210.

Brenowitz, E. A. 1981. The effect of stimulus presentation sequence on the response of red-winged blackbirds in playback studies. *Auk* 98:355–360.

Brenowitz, E. A. 1982a. Aggressive response of red-winged blackbirds to mockingbird song imitation. *Auk* 99:584–586.

Brenowitz, E. A. 1982b. Long-range communication of species identity by song in the red-winged blackbird. *Behav. Ecol. Sociobiol.* 10:29–38.

Brenowitz, E. A. 1983. The contribution of temporal song cues to species recognition in the red-winged blackbird. *Anim. Behav.* 31:1116–1127.

Brown, B. T., and J. W. Goertz. 1978. Reproduction and nest-site selection by red-winged blackbirds in north Louisiana. *Wilson Bull.* 90:261–270.

Brown, C. R. 1986. Cliff swallow colonies as information centers. *Science* 234:83–85.

Brown, J. L. 1964. The evolution of diversity in avian territorial systems. *Wilson Bull.* 6:160–169.

Brown, J. L. 1975. *The Evolution of Behavior.* Norton, New York.

Bryant, D. M. 1989. House martin. Pp. 89–106 in I. Newton, ed., *Lifetime Reproduction in Birds.* Academic Press, London.

Buitron, D. 1988. Female and male specialization in parental care and its consequences in black-billed magpies. *Condor* 90:29–39.

Bunzel, M., and J. Drüke. 1989. Kingfisher. Pp. 107–116 in I. Newton, ed., *Lifetime Reproduction in Birds.* Academic Press, London.

Burke, T., and M. W. Bruford. 1987. DNA fingerprinting in birds. *Nature* 327:149–152.

Burley, N. 1981. Mate choice by multiple criteria in a monogamous species. *Am. Nat.* 117:515–528.

Burley, N., G. Krantzberg, and P. Radman. 1982. Influence of colour-banding on the conspecific preferences of zebra finches. *Anim. Behav.* 30:444–455.

Burns, J. T. 1982. Nests, territories, and reproduction of sedge wrens (*Cistothorus platensis*). *Wilson Bull.* 94:338–349.

Caccamise, D. F. 1976. Nesting mortality in the red-winged blackbird. *Auk* 93:517–534.

Caccamise, D. F. 1978. Seasonal patterns of nesting mortality in the red-winged blackbird. *Condor* 80:290–294.

Calder, W. A., III. 1984. *Size, Function, and Life History.* Harvard University Press, Cambridge, Mass.

Capp, M. S. 1992. Tests of the function of the song repertoire in bobolinks. *Condor* 94:468–479.

Carey, M., and V. Nolan, Jr. 1979. Population dynamics of indigo buntings and the evolution of avian polygyny. *Evolution* 33:1180–1192.

Case, N. A., and O. H. Hewitt. 1963. Nesting and productivity of the red-winged blackbird in relation to habitat. *Living Bird* 2:7–20.

Catchpole, C. K. 1973. The functions of advertising song in the sedge warbler (*Acrocephalus schoenobaenus*) and the reed warbler (*A. scirpaceus*). *Behaviour* 46:300–320.

Catchpole, C. K. 1980. Sexual selection and the evolution of complex songs among European warblers of the genus *Acrocephalus*. *Behaviour* 74:149–166.

Catchpole, C. K. 1986. Song repertoires and reproductive success in the great reed warbler *Acrocephalus arundinaceus*. *Behav. Ecol. Sociobiol.* 19:439–445.

Catchpole, C. K., B. Leisler, and H. Winkler. 1985. Polygyny in the great reed warbler, *Acrocephalus arundinaceus*: a possible case of deception. *Behav. Ecol. Sociobiol.* 16:285–291.

Clutton-Brock, T. H. 1988. Reproductive success. Pp. 472–485 in T. H. Clutton-Brock, ed., *Reproductive Success: Studies of Individual Variation in Contrasting Breeding Systems.* University of Chicago Press, Chicago.

Clutton-Brock, T. H., P. H. Harvey, and B. Rudder. 1977. Sexual dimorphism, socionomic sex ratio and body weight in primates. *Nature* 269:797–799.

Clutton-Brock, T. H., and G. A. Parker. 1992. Potential reproductive rates and the operation of sexual selection. *Quart. Rev. Biol.* 67:437–456.

Clutton-Brock, T. H., and A. C. J. Vincent. 1991. Sexual selection and the potential reproductive rates of males and females. *Nature* 351:58–60.

Crawford, R. D. 1977a. Breeding biology of year-old and older female red-winged and yellow-headed blackbirds. *Wilson Bull.* 89:73–80.

Crawford, R. D. 1977b. Polygynous breeding of short-billed marsh wrens. *Auk* 94:359–362.

Cronmiller, J. R., and C. F. Thompson. 1980. Experimental manipulation of brood size in red-winged blackbirds. *Auk* 97:559–565.

Cronmiller, J. R., and C. F. Thompson. 1981. Sex-ratio adjustment in malnourished red-winged blackbird broods. *J. Field Ornithol.* 52:65–67.

Crow, J. F. 1958. Some possibilities for measuring selection intensities in man. *Human Biol.* 30:1–13.

Curio, E. 1973. Towards a methodology of teleonomy. *Experientia* 29:1045–1180.

Darwin, C. 1859. *On the Origin of Species by Means of Natural Selection.* John Murray, London.

Darwin, C. 1871. *The Descent of Man, and Selection in Relation to Sex.* John Murray, London.

Davies, N. B. 1978. Territorial defence in the speckled wood butterfly (*Parage aegeria*): the resident always wins. *Anim. Behav.* 26:138–147.

Davies, N. B. 1989. Sexual conflict and the polygamy threshold. *Anim. Behav.* 38:226–234.

Davies, N. B. 1992. *Dunnock Behaviour and Social Evolution.* Oxford University Press, Oxford.

Dawkins, R., and T. R. Carlisle. 1976. Parental investment, mate desertion and a fallacy. *Nature* 262:131–133.

Dawkins, R., and J. R. Krebs. 1978. Animal signals: information or manipulation? Pp. 282–309 in J. R. Krebs and N. B. Davies, eds., *Behavioural Ecology.* Sinauer, Sunderland, Mass.

Deevey, E. S., Jr. 1947. Life tables for natural populations of animals. *Quart. Rev. Biol.* 22:283–314.

Derrickson, K. C. 1987. Yearly and situational changes in the estimate of repertoire size in northern mockingbirds (*Mimus polyglottos*). *Auk* 104:198–207.

Dickinson, T. E. 1987. The vocal behavior of female red-winged blackbirds (*Agelaius phoeniceus*). Ph.D. diss., University of Pennsylvania, Philadelphia.

Dickinson, T. E., J. B. Falls, and J. Kopachena. 1987. Effects of female pairing status and timing of breeding on nesting productivity in western meadowlarks (*Sturnella neglecta*). *Can. J. Zool.* 65:3093–3101.

Dickinson, T. E., and M. R. Lein. 1987. Territory dynamics and patterns of female recruitment in red-winged blackbirds (*Agelaius phoeniceus*). *Can. J. Zool.* 65:465–471.

Dolbeer, R. A. 1976. Reproductive rate and temporal spacing of nesting of red-winged blackbirds in upland habitat. *Auk* 93:343–355.

Downhower, J. F. 1976. Darwin's finches and the evolution of sexual dimorphism in body size. *Nature* 263:558–563.

Downhower, J. F., and L. Brown. 1980. Mate preferences of female mottled sculpins, *Cottus bairdi. Anim. Behav.* 28:728–734.

Dugatkin, L. A. 1992. Sexual selection and imitation: females copy the mate choice of others. *Am. Nat.* 139:1384–1389.

Dyrcz, A. 1986. Factors affecting facultative polygyny and breeding results in the great reed warbler (*Acrocephalus arundinaceus*). *J. Ornithol.* 127:447–461.

Eckert, C. G., and P. J. Weatherhead. 1987a. Owners, floaters and competitive asymmetries among territorial red-winged blackbirds. *Anim. Behav.* 35:1317–1323.

Eckert, C. G., and P. J. Weatherhead. 1987b. Male characteristics, parental quality and the study of mate choice in the red-winged blackbird (*Agelaius phoeniceus*). *Behav. Ecol. Sociobiol.* 20:35–42.

Eckert, C. G., and P. J. Weatherhead. 1987c. Ideal dominance distributions: a test using red-winged blackbirds. *Behav. Ecol. Sociobiol.* 20:43–52.

Eckert, C. G., and P. J. Weatherhead. 1987d. Competition for territories in red-winged blackbirds: is resource-holding potential realized? *Behav. Ecol. Sociobiol.* 20:369–375.

Eliason, B. C. 1986. Female site fidelity and polygyny in the blackpoll warbler (*Dendroica striata*). *Auk* 103:782–790.

Emlen, S. T., and L. W. Oring. 1977. Ecology, sexual selection, and the evolution of mating systems. *Science* 197:215–223.

Endler, J. A. 1986. *Natural Selection in the Wild.* Princeton University Press, Princeton, N. J.

Erckmann, W. J., L. D. Beletsky, G. H. Orians, T. Johnsen, S. Sharbaugh, and C. D'Antonio. 1990. Old nests as cues for nest-site selection: an experimental test with red-winged blackbirds. *Condor* 92:113–117.

Ewald, P. W., and S. Rohwer. 1982. Effects of supplemental feeding on timing of breeding, clutch-size and polygyny in red-winged blackbirds *Agelaius phoeniceus*. *J. Anim. Ecol.* 51:429–450.

Ezaki, Y. 1990. Female choice and the causes and adaptiveness of polygyny in great reed warblers. *J. Anim. Ecol.* 59:103–119.

Facemire, C. F. 1980. Cowbird parasitism of marsh-dwelling red-winged blackbirds. *Condor* 82:347–348.

Falls, J. B. 1992. Playback: a historical perspective. Pp. 11–33 in P. K. McGregor, ed., *Playback and Studies of Animal Communication.* Plenum, New York.

Fiala, K. L. 1981. Reproductive cost and the sex ratio in red-winged blackbirds. Pp. 198–214 in R. D. Alexander and D. W. Tinkle, eds., *Natural Selection and Social Behavior.* Chiron Press, New York.

Fiala, K. L., and J. D. Congdon. 1983. Energetic consequences of sexual size dimorphism in nestling red-winged blackbirds. *Ecology* 64:642–647.

Ficken, M. S., and R. W. Ficken. 1970. Responses of four warbler species to playback of their two song types. *Auk* 87:296–304.

Fisher, R. A. 1930. *The Genetical Theory of Natural Selection.* Clarendon Press, Oxford.

Folstad, I., and A. J. Karter. 1992. Parasites, bright males, and the immunocompetence handicap. *Am. Nat.* 139:603–622.

Foster, M. S. 1987. Delayed maturation, neoteny, and social system differences in two manakins of the genus *Chiroxiphia*. *Evolution* 41:547–558.

Francis, W. J. 1975. Clutch size and nesting success in red-winged blackbirds. *Auk* 92:815–817.

Freeman, S. 1987. Male red-winged blackbirds (*Agelaius phoeniceus*) assess the RHP of neighbors by watching contests. *Behav. Ecol. Sociobiol.* 21:307–311.

Freeman, S. 1990. Molecular systematics and morphological evolution in the blackbirds. Ph.D. diss., University of Washington, Seattle.

Freeman, S., D. F. Gori, and S. Rohwer. 1990. Red-winged blackbirds and brown-headed cowbirds: some aspects of a host-parasite relationship. *Condor* 92:336–340.

Freeman, S., and W. M. Jackson. 1990. Univariate metrics are not adequate to measure avian body size. *Auk* 107:69–74.

Fretwell, S. D., and H. L. Lucas, Jr. 1969. On territorial behaviour and other factors influencing habitat distribution in birds. I. Theoretical development. *Acta Biotheor.* 19:16–36.

Friedmann, H. 1963. Host relations of the parasitic cowbirds. *U. S. Natl. Mus. Bull.* 233:1–276.

Garson, P. J., W. K. Pleszczynska, and C. H. Holm. 1981. The "polygyny threshold" model: a reassessment. *Can. J. Zool.* 59:902–910.

Gaulin, S. J. C., and L. D. Sailer. 1984. Sexual dimorphism in weight among the primates: the relative impact of allometry and sexual selection. *Int. J. Primatol.* 5:515–535.

Gavin, T. A., R. A. Howard, and B. May. 1991. Allozyme variation among breeding populations of red-winged blackbirds: the California conundrum. *Auk* 108: 602–611.

Ghiselin, M. T. 1974. *The Economy of Nature and the Evolution of Sex*. University of California Press, Berkeley.

Gibbs, H. L., P. J. Weatherhead, P. T. Boag, B. N. White, L. M. Tabak, and D. J. Hoysak. 1990. Realized reproductive success of polygynous red-winged blackbirds revealed by DNA markers. *Science* 250:1394–1397.

Gibson, R. M., and J. W. Bradbury. 1985. Sexual selection in lekking sage grouse: phenotypic correlates of male mating success. *Behav. Ecol. Sociobiol.* 18:117–123.

Gibson, R. M., J. W. Bradbury, and S. L. Vehrencamp. 1991. Mate choice in lekking sage grouse revisited: the roles of vocal display, female site fidelity, and copying. *Behav. Ecol.* 2:165–180.

Goddard, S. V., and V. V. Board. 1967. Reproductive success of red-winged blackbirds in north central Oklahoma. *Wilson Bull.* 79:283–289.

Grafen, A. 1987. The logic of divisively asymmetric contests: respect for ownership and the desperado effect. *Anim. Behav.* 35:462–467.

Grafen, A. 1988. On the uses of data on lifetime reproductive success. Pp. 454–471 in T. H. Clutton-Brock, ed., *Reproductive Success: Studies of Individual Variation in Contrasting Breeding Systems*. University of Chicago Press, Chicago.

Greene, E. 1987. Individuals in an osprey colony discriminate between high and low quality information. *Nature* 329:239–241.

Hagan, J. M., and J. M. Reed. 1988. Red color bands reduce fledging success in red-cockaded woodpeckers. *Auk* 105:498–503.

Haigh, C. R. 1968. Sexual dimorphism, sex ratios, and polygyny in the red-winged blackbird. Ph.D. diss., University of Washington, Seattle.

Hamilton, W. D., and M. Zuk. 1982. Heritable true fitness and bright birds: a role for parasites? *Science* 218:384–387.

Hansen, A. J., and S. Rohwer. 1986. Coverable badges and resource defence in birds. *Anim. Behav.* 34:69–76.

Harding, C. F., M. J. Walters, D. Collado, and K. Sheridan. 1988. Hormonal specificity and activation of social behavior in male red-winged blackbirds. *Horm. Behav.* 22:402–418.

Harms, K. E., L. D. Beletsky, and G. H. Orians. 1991. Conspecific nest parasitism in three species of New World blackbirds. *Condor* 93:967–974.

Hastings, P. A. 1988. Female choice and male reproductive success in the angel blenny, *Coralliozetus angelica* (Teleostei: Chaenopsidae). *Anim. Behav.* 36:115–124.

Heagy, P. A., and L. B. Best. 1983. Factors affecting feeding and brooding of brown thrasher nestlings. *Wilson Bull.* 95:297–303.

Hill, G. E. 1990. Female house finches prefer colourful males: sexual selection for a condition-dependent trait. *Anim. Behav.* 40:563–572.

Hill, G. E. 1992. Proximate basis of variation in carotenoid pigmentation in male house finches. *Auk* 109:1–12.

Hinde, R. A. 1970. Behavioural habituation. Pp. 3–40 in G. Horn and R. A. Hinde, eds., *Short-term Changes in Neural Activity and Behaviour.* Cambridge University Press, Cambridge.

Holcomb, L. C. 1974. The question of possible surplus females in breeding red-winged blackbirds. *Auk* 86:177–179.

Holcomb, L. C., and G. Tweist. 1968. Ecological factors affecting nest building in red-winged blackbirds. *Bird-Banding* 39:14–22.

Holm, C. H. 1973. Breeding sex ratios, territoriality, and reproductive success in the red-winged blackbird (*Agelaius phoeniceus*). *Ecology* 54:356–365.

Hötker, H. 1989. Meadow pipit. Pp. 119–133 in I. Newton, ed., *Lifetime Reproduction in Birds*. Academic Press, London.

Houde, A. E. 1987. Mate choice based upon naturally occurring color-pattern variation in a guppy population. *Evolution* 41:1–10.

Howard, E. 1920. *Territory in Bird Life*. Collins Sons and Co., London.

Howard, R. D. 1980. Mating behaviour and mating success in woodfrogs, *Rana sylvatica. Anim. Behav.* 28:705–716.

Howard, R. D. 1981. Male age-size distribution and male mating success in bullfrogs. Pp. 61–77 in R. D. Alexander and D. W. Tinkle, eds., *Natural Selection and Social Behavior*. Chiron Press, New York.

Howe, H. F. 1979. Evolutionary aspects of parental care in the common grackle, *Quiscalus quiscula* L. *Evolution* 33:41–51.

Hurd, P. L., P. J. Weatherhead, and S. B. McRae. 1991. Parental consumption of nestling feces: good food or sound economics? *Behav. Ecol.* 2:69–76.

Hurlbert, S. H. 1984. Pseudoreplication and the design of ecological field experiments. *Ecol. Monogr.* 54:187–211.

Hurly, T. A., and R. J. Robertson. 1984. Aggressive and territorial behaviour in female red-winged blackbirds. *Can. J. Zool.* 62:148–153.

Hurly, T. A., and R. J. Robertson. 1985. Do female red-winged blackbirds limit harem size? I. A removal experiment. *Auk* 102:205–209.

Huxley, J. S. 1938. The present standing of the theory of sexual selection. Pp. 11–42 in G. DeBeer, ed., *Evolution: Essays on Aspects of Evolutionary Biology*. Clarendon Press, Oxford.

Irwin, R. E. 1990. Directional sexual selection cannot explain variation in song repertoire size in the New World blackbirds (Icterinae). *Ethology* 85:212–224.

Janetos, A. C. 1980. Strategies of mate choice: a theoretical analysis. *Behav. Ecol. Sociobiol.* 7:107–112.

Jarman, M. V. 1979. Impala social behaviour: territory, hierarchy, mating, and the use of space. *Adv. Ecol. Res.* 21:1–92.

Jeffreys, A. J., V. Wilson, and S. L. Thein. 1985. Hypervariable "minisatellite" regions in human DNA. *Nature* 314:67–73.

Johnsen, T. S. 1991. Steriod hormones and male reproductive behavior in red-winged blackbirds (*Agelaius phoeniceus*): seasonal variation and behavioral correlates of testosterone. Ph.D. diss., Indiana University, Bloomington.

Johnson, D. M., G. L. Stewart, M. Corley, R. Ghrist, J. Hagner, A. Ketterer, B. McDonnell, W. Newsom, E. Owen, and P. Samuels. 1980. Brown-headed cowbird (*Molothrus ater*) mortality in an urban winter roost. *Auk* 97:299–320.

Johnson, L. S., and L. H. Kermott. 1991. Effect of nest-site supplementation on polygynous behavior in the house wren (*Troglodytes aedon*). *Condor* 93:784–787.

Johnson, L. S., and L. H. Kermott. 1993. Why is reduced male parental assistance detrimental to the reproductive success of secondary female house wrens? *Anim. Behav.* 46:1111–1120.

Johnson, L. S., L. H. Kermott, and M. R. Lein. 1993. The cost of polygyny in the house wren *Troglodytes aedon*. *J. Anim. Ecol.* 62:669–682.

Kirkpatrick, M. 1982. Sexual selection and the evolution of female choice. *Evolution* 36:1–12.

Kirkpatrick, M. 1987. Sexual selection by female choice in polygynous animals. *Ann. Rev. Ecol. Syst.* 18:43–70.

Kirkpatrick, M., and M. J. Ryan. 1991. The evolution of mating preferences and the paradox of the lek. *Nature* 350:33–38.

Kitchen, D. W. 1974. Social behavior and ecology of the pronghorn. *Wild. Monogr.* 38:1–96.

Knight, R. L., S. Kim, and S. A. Temple. 1985. Predation of red-winged blackbird nests by mink. *Condor* 87:304–305.

Knight, R. L., and S. A. Temple. 1986. Why does intensity of avian nest defense increase during the nesting cycle? *Auk* 103:318–327.

Knight, R. L., and S. A. Temple. 1988. Nest-defense behavior in the red-winged blackbird. *Condor* 90:193–200.

Kodric-Brown, A. 1983. Determinants of male reproductive success in pupfish (*Cyprinodon pecosensis*). *Anim. Behav.* 31:128–137.

Korpimäki, E. 1991. Poor reproductive success of polygynously mated female Tengmalm's owls: are better options available? *Anim. Behav.* 41:37–47.

Krebs, J. R. 1977. The significance of song repertoires: the Beau Geste hypothesis. *Anim. Behav.* 25:475–478.

Krebs, J. R., R. Ashcroft, and M. I. Webber. 1978. Song repertoires and territory defence in the great tit. *Nature* 271:539–542.

Krebs, J. R., and D. E. Kroodsma. 1980. Repertoires and geographic variation in bird song. *Adv. Study Behav.* 11:143–178.

Krebs, R. A., and D. A. West. 1988. Female mate preference and the evolution of female-limited Batesian mimicry. *Evolution* 42:1101–1104.

Kroodsma, D. E., and F. C. James. 1994. Song variation within and among populations of red-winged blackbirds. *Wilson Bull.* in press.

Lack, D. 1968. *Ecological Adaptations for Breeding in Birds*. Methuen, London.

Lack, D., and J. T. Emlen. 1939. Observations on breeding behavior in tricolored redwings. *Condor* 41:225–230.

Lande, R. 1980. Sexual dimorphism, sexual selection, and adaptation in polygenic characters. *Evolution* 34:292–305.

Langston, N. E., S. Freeman, S. Rohwer, and D. Gori. 1990. The evolution of female body size in red-winged blackbirds: the effects of timing of breeding, social competition, and reproductive energetics. *Evolution* 44:1764–1779.

LaPrade, H. R., and H. B. Graves. 1982. Polygyny and female-female aggression in red-winged blackbirds (*Agelaius phoeniceus*). *Am. Nat.* 120:135–138.

Leffelaar, D., and R. J. Robertson. 1986. Equality of feeding roles and the maintenance of monogamy in tree swallows. *Behav. Ecol. Sociobiol.* 18:199–206.

Lenington, S. 1977. Evolution of polygyny in redwinged blackbirds. Ph.D. diss., University of Chicago, Chicago.

Lenington, S. 1980. Female choice and polygyny in redwinged blackbirds. *Anim. Behav.* 28:347–361.

Leonard, M. L. 1990. Polygyny in marsh wrens: asynchronous settlement as an alternative to the polygyny-threshold model. *Am. Nat.* 136:446–458.

Leonard, M. L., and J. Picman. 1987. Female settlement in marsh wrens: is it affected by other females? *Behav. Ecol. Sociobiol.* 21:135–140.

Leonard, M. L., and J. Picman. 1988. Mate choice by marsh wrens: the influence of male and territory quality. *Anim. Behav.* 36:517–528.

Lifjeld, J. T., and T. Slagsvold. 1989. Allocation of parental investment by polygynous pied flycatcher males. *Ornis Fennica* 66:3–14.

Lightbody, J. P., and P. J. Weatherhead. 1987a. Polygyny in the yellow-headed blackbird: female choice versus male competition. *Anim. Behav.* 35:1670–1684.

Lightbody, J. P., and P. J. Weatherhead. 1987b. Interactions among females in polygynous yellow-headed blackbirds. *Behav. Ecol. Sociobiol.* 21:23–30.

Lightbody, J. P., and P. J. Weatherhead. 1988. Female settling patterns and polygyny: tests of a neutral-mate-choice hypothesis. *Am. Nat.* 132:20–33.

Linz, G. M., and S. B. Bolin. 1982. Incidence of brown-headed cowbird parasitism on red-winged blackbirds. *Wilson Bull.* 94:93–95.

Lyon, B. E., and R. D. Montgomerie. 1986. Delayed plumage maturation in passerine birds: reliable signaling by subordinate males? *Evolution* 40:605–615.

Martin, S. G. 1971. Polygyny in the bobolink: habitat quality and the adaptive complex. Ph.D. diss., Oregon State Univ., Corvallis.

Maynard Smith, J. 1977. Parental investment: a prospective analysis. *Anim. Behav.* 25:1–9.

Maynard Smith, J. 1979. Game theory and the evolution of behaviour. *Proc. R. Soc. Lond. B*. 205:475–488.

Maynard Smith, J. 1991. Theories of sexual selection. *Trends Ecol. Evol.* 6:146–151.

Maynard Smith, J., and G. A. Parker. 1976. The logic of asymmetric contests. *Anim. Behav.* 24:159–175.

McDonald, M. V. 1989. Function of song in Scott's seaside sparrow, *Ammodramus maritimus peninsulae*. *Anim. Behav.* 38:468–465.

McGillivray, W. B. 1984. Nestling feeding rates and body size of adult house sparrows. *Can. J. Zool.* 62:381–385.

Meanley, B. 1964. Origin, structure, molt, and dispersal of a late summer red-winged blackbird population. *Bird-Banding* 35:32–38.

Meanley, B., and J. S. Webb. 1963. Nesting ecology and reproductive rate of the red-winged blackbird in tidal marshes of the upper Chesapeake Bay region. *Chesapeake Sci.* 4:90–100.

Metz, K. J., and P. J. Weatherhead. 1991. Color bands function as secondary sexual traits in male red-winged blackbirds. *Behav. Ecol. Sociobiol.* 28:23–27.

Metz, K. J., and P. J. Weatherhead. 1992. Seeing red: uncovering coverable badges in red-winged blackbirds. *Anim. Behav.* 43:223–229.

Miller, R. S. 1968. Conditions of competition between redwings and yellowheaded blackbirds. *J. Anim. Ecol.* 37:43–62.

Mills, J. A. 1989. Red-billed gull. Pp. 387–404 in I. Newton, ed., *Lifetime Reproduction in Birds*. Academic Press, London.

Miskimen, M. 1980. Red-winged blackbirds. I. Age-related epaulet color changes in captive females. *Ohio J. Sci.* 80:232–235.

Møller, A. P. 1988. Badge size in the house sparrow *Passer domesticus*: Effects of intra- and intersexual selection. *Behav. Ecol. Sociobiol.* 22:373–378.

Monnett, C., L. M. Rotterman, C. Worlein, and K. Halupka. 1984. Copulation patterns of red-winged blackbirds (*Agelaius phoeniceus*). *Am. Nat.* 124:757–764.

Morris, L. 1975. Effect of blackened epaulets on the territorial behavior and breeding success of male redwinged blackbirds, *Agelaius phoeniceus*. *Ohio J. Sci.* 75:168–176.

Morse, D. H. 1966. The contexts of songs in the yellow warbler. *Wilson Bull.* 78:444–455.

Morse, D. H. 1970. Territorial and courtship songs of birds. *Nature* 226:659–661.

Muldal, A. M., J. D. Moffatt, and R. J. Robertson. 1986. Parental care of nestlings by male red-winged blackbirds. *Behav. Ecol. Sociobiol.* 19:105–114.

Muma, K. E., and P. J. Weatherhead. 1989. Male traits expressed in females: direct or indirect sexual selection? *Behav. Ecol. Sociobiol.* 25:23–31.

Muma, K. E., and P. J. Weatherhead. 1991. Plumage variation and dominance in captive female red-winged blackbirds. *Can. J. Zool.* 69:49–54.

Munsell, A. H. 1961. *A Color Notation*. 11th edn. Munsell Color Co., Baltimore.

Nero, R. W. 1956a. A behavior study of the red-winged blackbird. I. Mating and nesting activities. *Wilson Bull.* 68:5–37.

Nero, R. W. 1956b. A behavior study of the red-winged blackbird. II. Territoriality. *Wilson Bull.* 68:129–150.

Nero, R. W. 1984. *Redwings*. Smithsonian Institution Press, Washington, D. C.

Noble, G. K. 1939. The role of dominance in the social life of birds. *Auk* 56:263–273.

Nolan, V., Jr. 1978. The ecology and behavior of the prairie warbler *Dendroica discolor*. *Ornithol. Monogr.* 26:1–595.

O'Donald, P. 1980. *Genetic Models of Sexual Selection*. Cambridge University Press, Cambridge.

Orians, G. H. 1961. The ecology of blackbird (*Agelaius*) social systems. *Ecol. Monogr.* 31:285–312.

Orians, G. H. 1969a. On the evolution of mating systems in birds and mammals. *Am. Nat.* 103:589–603.

Orians, G. H. 1969b. Age and hunting success in the brown pelican (*Pelecanus occidentalis*). *Anim. Behav.* 17:316–319.

Orians, G. H. 1972. The adaptive significance of mating systems in the Icteridae. *Proc. XV Intern. Ornithol. Congr.* 389–398.

Orians, G. H. 1973. The red-winged blackbird in tropical marshes. *Condor* 75:28–42.

Orians, G. H. 1980. *Some Adaptations of Marsh-nesting Blackbirds*. Princeton University Press, Princeton, N. J.

Orians, G. H. 1985. *Blackbirds of the Americas*. University of Washington Press, Seattle.

Orians, G. H., and L. D. Beletsky. 1989. Red-winged blackbird. Pp. 183–197 in I. Newton, ed., *Lifetime Reproduction in Birds*. Academic Press, London.

Orians, G. H., and G. M. Christman. 1968. A comparative study of the behavior of red-winged, tricolored, and yellow-headed blackbirds. *Univ. Calif. Publ. Zool.* 84:1–81.

Orians, G. H., and G. Collier. 1963. Competition and blackbird social systems. *Evolution* 17:449–459.

Orians, G. H., and M. F. Willson. 1964. Interspecific territories of birds. *Ecology* 45:736–745.

Ortega, C. P., and A. Cruz. 1988. Mechanisms of egg acceptance by marsh-dwelling blackbirds. *Condor* 90:349–358.

Parker, G. A. 1974. Assessment strategy and the evolution of fighting behaviour. *J. Theor. Biol.* 47:223–243.

Patterson, C. B. 1979. Relative parental investment in the red-winged blackbird. Ph.D. diss., Indiana University, Bloomington.

Patterson, C. B. 1991. Relative parental investment in the red-winged blackbird. *J. Field Ornithol.* 62:1–18.

Patterson, C. B., W. J. Erckmann, and G. H. Orians. 1980. An experimental study of parental investment and polygyny in male blackbirds. *Am. Nat.* 116:757–769.

Payne, R. B. 1969. Breeding seasons and reproductive physiology of tricolored blackbirds and redwinged blackbirds. *Univ. Calif. Publ. Zool.* 90:1–115.

Payne, R. B. 1979. Sexual selection and intersexual differences in variance of breeding success. *Am. Nat.* 114:447–452.

Payne, R. B. 1984. Sexual selection, lek and arena behavior and sexual size dimorphism in birds. *Ornithol. Monogr.* 33:1–52.

Peek, F. W. 1971. Seasonal change in the breeding behavior of the male red-winged blackbird. *Wilson Bull.* 83:383–395.

Peek, F. W. 1972. An experimental study of the territorial function of vocal and visual display in the male red-winged blackbird (*Agelaius phoeniceus*). *Anim. Behav.* 20:112–118.

Perrone, M., Jr. 1975. The relation between mate choice and parental investment patterns in fish who brood their young: theory and case study. Ph.D. diss., University of Washington, Seattle.

Petit, L. J. 1991. Experimentally induced polygyny in a monogamous bird species: prothonotary warblers and the polygyny threshold. *Behav. Ecol. Sociobiol.* 29:177–187.

Pianka, E. R. 1975. Niche relations of desert lizards. Pp. 292–314 in M. L. Cody and J. M. Diamond, eds., *Ecology and Evolution of Communities.* Harvard University Press, Cambridge, Mass.

Picman, J. 1977a. Destruction of eggs by the long-billed marsh wren (*Telmatodytes palustris palustris*). *Can. J. Zool.* 55:1914–1920.

Picman, J. 1977b. Intraspecific nest destruction in the long-billed marsh wren, *Telmatodytes palustris palustris*. *Can. J. Zool.* 55:1997–2003.

Picman, J. 1980a. Impact of marsh wrens on reproductive strategy of red-winged blackbirds. *Can. J. Zool.* 58:337–350.

Picman, J. 1980b. Behavioural interactions between red-winged blackbirds and long-billed marsh wrens and their role in the evolution of the redwing polygynous mating system. Ph.D. diss., University of British Columbia, Vancouver.

Picman, J. 1981. The adaptive value of polygyny in marsh-nesting red-winged blackbirds; renesting, territory tenacity, and mate fidelity of females. *Can. J. Zool.* 59:2284–2296.

Picman, J. 1987. Territory establishment, size, and tenacity by male red-winged blackbirds. *Auk* 104:405–412.

Picman, J., M. Leonard, and A. Horn. 1988. Antipredation role of clumped nesting by marsh-nesting red-winged blackbirds. *Behav. Ecol. Sociobiol.* 22:9–15.

Picman, J., M. L. Milks, and M. Leptich. 1994. Patterns of predation on passerine nests in marshes: effects of water depth and distance from edge. *Auk* in press.

Pitelka, F. A. 1959. Numbers, breeding schedule, and territoriality in pectoral sandpipers in northern Alaska. *Condor* 61:233–264.

Platt, J. R. 1964. Strong inference. *Science* 146:347–353.

Post, W. 1981. Biology of the yellow-shouldered blackbird—*Agelaius* on a tropical island. *Bull. Florida State Museum* 26:125–202.

Postupalsky, S. 1989. Osprey. Pp. 297–313 in I. Newton, ed., *Lifetime Reproduction in Birds.* Academic Press, London.

Pruett-Jones, S. 1992. Independent versus nonindependent mate choice: do females copy each other? *Am. Nat.* 140:1000–1009.

Radesäter, T., S. Jakobsson, N. Andbjer, A. Bylin, and K. Nyström. 1987. Song rate and pair formation in the willow warbler, *Phylloscopus trochilus*. *Anim. Behav.* 35:1645–1651.

Real, L. 1990. Search theory and mate choice. I. Models of single-sex discrimination. *Am. Nat.* 136:376–404.

Recher, H. F., and J. A. Recher. 1969. Comparative foraging efficiency of adult and immature little blue herons (*Florida caerulea*). *Anim. Behav.* 17:320–322.

Ricklefs, R. E. 1973. Fecundity, mortality, and avian demography. Pp. 366–435 in D. S. Farner, ed., *Breeding Biology of Birds*. Natl. Acad. Sciences, Washington, D. C.

Ricklefs, R. E. 1974. Energetics of reproduction in birds. Pp. 152–297 in R. A. Paynter, Jr., ed., *Avian Energetics*. Publ. Nuttall Ornithological Club, No. 15. Cambridge, Mass.

Ridley, M. 1983. *The Explanation of Organic Diversity: The Comparative Method and Adaptations for Mating*. Clarendon Press, Oxford.

Rising, J. D., and K. M. Somers. 1989. The measurement of overall body size in birds. *Auk* 106:666–674.

Ritschel, S. E. 1985. Breeding ecology of the red-winged blackbird (*Agelaius phoeniceus*): tests of models of polygyny. Ph.D. diss., University of California, Irvine.

Roberts, L. B., and W. A. Searcy. 1988. Dominance relationships in harems of female red-winged blackbirds. *Auk* 105:89–96.

Robertson, R. J. 1972. Optimal niche space of the redwinged blackbird (*Agelaius phoeniceus*). I. Nesting success in marsh and upland habitat. *Can. J. Zool.* 50:247–263.

Robertson, R. J. 1973. Optimal niche space of the redwinged blackbird: spatial and temporal patterns of nesting activity and success. *Ecology* 54:1085–1093.

Robertson, R. J., and R. F. Norman. 1976. Behavioral defenses to brood parasitism by potential hosts of the brown-headed cowbird. *Condor* 78:166–173.

Robinson, S. K. 1986. Benefits, costs, and determinants of dominance in a polygynous oriole. *Anim. Behav.* 34:241–255.

Rohwer, S. 1975. The social significance of avian winter plumage variability. *Evolution* 29:593–610.

Rohwer, S. 1978. Passerine subadult plumages and the deceptive acquisition of resources: test of a critical assumption. *Condor* 80:173–179.

Rohwer, S. 1982. The evolution of reliable and unreliable badges of fighting ability. *Am. Zool.* 22:531–546.

Rohwer, S., and G. S. Butcher. 1988. Winter versus summer explanations of delayed plumage maturation in temperate passerine birds. *Am. Nat.* 131:556–572.

Rohwer, S., S. D. Fretwell, and D. M. Niles. 1980. Delayed maturation in passerine plumages and the deceptive acquisition of resources. *Am. Nat.* 115:400–437.

Rohwer, S., and E. Røskaft. 1989. Results of dyeing male yellow-headed blackbirds solid black: implications for the arbitrary identity badge hypothesis. *Behav. Ecol. Sociobiol.* 25:39–48.

Røskaft, E., and S. Rohwer. 1987. An experimental study of the function of the red epaulettes and the black body colour of male red-winged blackbirds. *Anim. Behav.* 35:1070–1077.

Rothstein, S. I. 1975. An experimental and telenomic investigation of avian brood parasitism. *Condor* 77:250–271.

Rowley, I., and E. Russell. 1989. Splendid fairy-wren. Pp. 233–252 in I. Newton, ed., *Lifetime Reproduction in Birds*. Academic Press, London.

Ruby, D. E. 1981. Phenotypic correlates of male reproductive success in the lizard *Sceloperus jarrovi*. Pp. 96–107 in R. D. Alexander and D. W. Tinkle, eds., *Natural Selection and Social Behavior*. Chiron Press, New York.

Ryan, M. J., J. H. Fox, W. Wilczynski, and A. S. Rand. 1990. Sexual selection for sensory exploitation in the frog *Physalaemus pustulosus*. *Nature* 343:66–67.

Ryves, B. H., and H. H. Ryves. 1934a. The breeding-habits of the corn-bunting as observed in North Cornwall: with special reference to its polygamous habit. *Brit. Birds* 28:2–26.

Ryves, B. H., and H. H. Ryves. 1934b. Supplementary notes on the breeding-habits of the corn-bunting as observed in North Cornwall in 1934. *Brit. Birds* 28:154–164.

Schoener, T. W. 1968. Size of feeding territories among birds. *Ecology* 49:123–141.

Searcy, W. A. 1979a. Female choice of mates: a general model for birds and its application to red-winged blackbirds (*Agelaius phoeniceus*). *Am. Nat.* 114:77–100.

Searcy, W. A. 1979b. Male characteristics and pairing success in red-winged blackbirds. *Auk* 96:353–363.

Searcy, W. A. 1979c. Sexual selection and body size in male red-winged blackbirds. *Evolution* 33:649–661.

Searcy, W. A. 1979d. Morphological correlates of dominance in captive male red-winged blackbirds. *Condor* 81:417–420.

Searcy, W. A. 1981. Sexual selection and aggressiveness in male red-winged blackbirds. *Anim. Behav.* 29:958–960.

Searcy, W. A. 1982. The evolutionary effects of mate selection. *Ann. Rev. Ecol. Syst.* 13:57–85.

Searcy, W. A. 1986. Are female red-winged blackbirds territorial? *Anim. Behav.* 34:1381–1391.

Searcy, W. A. 1988a. Do female red-winged blackbirds limit their own breeding densities? *Ecology* 69:85–95.

Searcy, W. A. 1988b. Dual intersexual and intrasexual functions of song in red-winged blackbirds. *Proc. XIX Intern. Ornithol. Congr.* 1:1373–1381.

Searcy, W. A. 1989. Function of male courtship vocalizations in red-winged blackbirds. *Behav. Ecol. Sociobiol.* 24:325–331.

Searcy, W. A. 1990. Species recognition of song by female red-winged blackbirds. *Anim. Behav.* 40:1119–1127.

Searcy, W. A. 1992. Song repertoire and mate choice in birds. *Am. Zool.* 32:71–80.

Searcy, W. A., and M. Andersson. 1986. Sexual selection and the evolution of song. *Ann. Rev. Ecol. Syst.* 17:507–533.

Searcy, W. A., and E. A. Brenowitz. 1988. Sexual differences in species recognition of avian song. *Nature* 332:152–154.

Searcy, W. A., S. Coffman, and D. F. Raikow. In press. Habituation, recovery, and the similarity of song types within repertoires in red-winged blackbirds (*Agelaius phoeniceus*) (Aves, Emberizidae). *Ethology*.

Searcy, W. A., D. Eriksson, and A. Lundberg. 1991. Deceptive behavior in pied flycatchers. *Behav. Ecol. Sociobiol.* 29:167–175.

Searcy, W. A., M. H. Searcy, and P. Marler. 1982. The response of swamp sparrows to acoustically distinct song types. *Behaviour* 80:70–83.

Searcy, W. A., and J. C. Wingfield. 1980. The effects of androgen and antiandrogen on dominance and aggressiveness in male red-winged blackbirds. *Horm. Behav.* 14:126–135.

Searcy, W. A., and K. Yasukawa. 1981a. Sexual size dimorphism and survival of male and female blackbirds (Icteridae). *Auk* 98:457–465.

Searcy, W. A., and K. Yasukawa. 1981b. Does the "sexy son" hypothesis apply to mate choice in red-winged blackbirds? *Am. Nat.* 117:343–348.

Searcy, W. A., and K. Yasukawa. 1983. Sexual selection and red-winged blackbirds. *Am. Sci.* 71:166–174.

Searcy, W. A., and K. Yasukawa. 1989. Alternative models of territorial polygyny in birds. *Am. Nat.* 134:323–343.

Searcy, W. A., and K. Yasukawa. 1990. Use of the song repertoire in intersexual and intrasexual contexts by male red-winged blackbirds. *Behav. Ecol. Sociobiol.* 27:123–128.

Selander, R. K. 1958. Age determination and molt in the boat-tailed grackle. *Condor* 60:355–376.

Selander, R. K. 1965. On mating systems and sexual selection. *Am. Nat.* 99:129–141.

Selander, R. K. 1972. Sexual selection and dimorphism in birds. Pp. 180–230 in B. Campbell, ed., *Sexual Selection and the Descent of Man 1871–1971.* Aldine, Chicago.

Shutler, D., and P. J. Weatherhead. 1991a. Owner and floater red-winged blackbirds: determinants of status. *Behav. Ecol. Sociobiol.* 28:235–241.

Shutler, D., and P. J. Weatherhead. 1991b. Basal song rate variation in male red-winged blackbirds: sound and fury signifying nothing? *Behav. Ecol.* 2:123–132.

Shutler, D., and P. J. Weatherhead. 1992. Surplus territory contenders in male red-winged blackbirds: where are the desperados? *Behav. Ecol. Sociobiol.* 31:97–106.

Simmons, R. E., P. C. Smith, and R. B. MacWhirter. 1986. Hierarchies among northern harrier (*Circus cyaneus*) harems and the costs of polygyny. *J. Anim. Ecol.* 55:755–771.

Sinclair, D. F. 1985. On tests of spatial randomness using nearest neighbor distances. *Ecology* 66:1084–1085.

Skutch, A. F. 1935. Helpers at the nest. *Auk* 52:257–273.

Small, M. P., and P. D. Boersma. 1990. Why female red-winged blackbirds call at the nest. *Wilson Bull.* 102:154–160.

Smith, D. G. 1972. The role of the epaulets in the red-winged blackbird (*Agelaius phoeniceus*) social system. *Behaviour* 41:251–268.

Smith, D. G. 1976. An experimental analysis of the function of red-winged blackbird song. *Behaviour* 56:136–156.

Smith, D. G. 1979. Male singing ability and territory integrity in red-winged blackbirds (*Agelaius phoeniceus*). *Behaviour* 68:193–206.

Smith, D. G., and F. A. Reid. 1979. Roles of the song repertoire in red-winged blackbirds. *Behav. Ecol. Sociobiol.* 5:279–290.

Smith, H. G., H. Källander, K. Fontell, and M. Ljungström. 1988. Feeding frequency and parental division of labour in the double-brooded great tit *Parus major*: effects of manipulating brood size. *Behav. Ecol. Sociobiol.* 22:447–453.

Smith, H. M. 1943. Size of breeding populations in relation to egg-laying and reproductive success in the eastern red-wing (*Agelaius p. phoeniceus*). *Ecology* 24: 183–207.

Smithe, F. B. 1975. *Naturalist's Color Guide*. American Museum of Natural History; New York.

Snedden, W. A., and S. K. Sakaluk. 1992. Acoustic signalling and its relation to male mating success in sagebrush crickets. *Anim. Behav.* 44:633–639.

Snelling, J. C. 1968. Overlap in feeding habits of redwinged blackbirds and common grackles nesting in a cattail marsh. *Auk* 85:560–585.

Spinage, C. A. 1982. *A Territorial Antelope: The Uganda Waterbuck*. Academic Press, London.

Stenmark, G., T. Slagsvold, and J. T. Lifjeld. 1988. Polygyny in the pied flycatcher, *Ficedula hypoleuca*: a test of the deception hypothesis. *Anim. Behav.* 36:1646–1657.

Sternberg, H. 1989. Pied flycatcher. Pp. 55–74 in I. Newton, ed., *Lifetime Reproduction in Birds*. Academic Press, London.

Strehl, C. E., and J. White. 1986. Effects of superabundant food on breeding success and behavior of the red-winged blackbird. *Oecologia* 70:178–186.

Teather, K. L. 1992. An experimental study of competition for food between male and female nestlings of the red-winged blackbird. *Behav. Ecol. Sociobiol.* 31:81–87.

Teather, K. L., K. E. Muma, and P. J. Weatherhead. 1988. Estimating female settlement from nesting data. *Auk* 105:196–200.

Temrin, H., and S. Jakobsson. 1988. Female reproductive success and nest predation in polyterritorial wood warblers (*Phylloscopus sibilatrix*). *Behav. Ecol. Sociobiol.* 23:225–231.

Thompson, C. W., and M. C. Moore. 1991. Throat colour reliably signals status in male tree lizards, *Urosaurus ornatus*. *Anim. Behav.* 42:745–753.

Trivers, R. L. 1972. Parental investment and sexual selection. Pp. 136–179 in B. Campbell, ed., *Sexual Selection and the Descent of Man 1871–1971*. Aldine, Chicago.

Trivers, R. L., and D. E. Willard. 1973. Natural selection of parental ability to vary the sex ratio of offspring. *Science* 179:90–92.

Ueda, K. 1984. Successive nest building and polygyny of fan-tailed warblers *Cisticola juncidis*. *Ibis* 126:221–229.

Urano, E. 1990. Factors affecting the cost of polygynous breeding for female great reed warblers *Acrocephalus arundinaceus*. *Ibis* 132:584–594.

Veiga, J. P. 1990. Infanticide by male and female house sparrows. *Anim. Behav.* 39:496–502.

Verner, J. 1964. Evolution of polygamy in the long-billed marsh wren. *Evolution* 18:252–261.

Verner, J., and G. H. Engelsen. 1970. Territories, multiple nest building, and polygyny in the long-billed marsh wren. *Auk* 87:557–567.

Verner, J., and M. F. Willson. 1966. The influence of habitats on mating systems of North American passerine birds. *Ecology* 47:143–147.

Wade, M. J. 1979. Sexual selection and variance in reproductive success. *Am. Nat.* 114:742–746.

Wade, M. J., and S. J. Arnold. 1980. The intensity of sexual selection in relation to male sexual behaviour, female choice, and sperm precedence. *Anim. Behav.* 28:446–461.

Wallace, A. R. 1889. *Darwinism: An Exposition of the Theory of Natural Selection with Some of Its Applications.* Macmillan and Co., London.

Walther, F. R., E. C. Mungall, and G. A. Grau. 1983. *Gazelles and Their Relatives: A study in Territorial Behavior.* Noyes, Park Ridge, N. J.

Ward, P., and A. Zahavi. 1973. The importance of certain assemblages of birds as "information-centres" for food-finding. *Ibis* 115:517–534.

Waser, P. M., and R. H. Wiley. 1979. Mechanisms and evolution of spacing in animals. Pp. 159–223 in P. Marler and J. G. Vandenbergh, eds., *Social Behavior and Communication.* Plenum, New York.

Wasserman, F. E. 1977. Mate attraction function of song in the white-throated sparrow. *Condor* 79:125–127.

Weatherhead, P. J. 1983. Secondary sex ratio adjustment in red-winged blackbirds (*Agelaius phoeniceus*). *Behav. Ecol. Sociobiol.* 12:57–61.

Weatherhead, P. J. 1984. Mate choice in avian polygyny: why do females prefer older males? *Am. Nat.* 123:873–875.

Weatherhead, P. J. 1985. Sex ratios of red-winged blackbirds by egg size and laying sequence. *Auk* 102:298–304.

Weatherhead, P. J. 1989. Sex ratios, host-specific reproductive success, and impact of brown-headed cowbirds. *Auk* 106:358–366.

Weatherhead, P. J. 1990a. Nest defence as shareable paternal care in red-winged blackbirds. *Anim. Behav.* 39:1173–1178.

Weatherhead, P. J. 1990b. Secondary sexual traits, parasites, and polygyny in red-winged blackbirds, *Agelaius phoeniceus. Behav. Ecol.* 1:125–130.

Weatherhead, P. J., and R. G. Clark. 1994. Natural selection and sexual size dimorphism in red-winged blackbirds. *Evolution* in press.

Weatherhead, P. J., H. Greenwood, and R. G. Clark. 1984. On the use of avian mortality patterns to test sexual selection theory. *Auk* 101:134–139.

Weatherhead, P. J., H. Greenwood, and R. G. Clark. 1987. Natural selection and sexual selection on body size in red-winged blackbirds. *Evolution* 41:1401–1403.

Weatherhead, P. J., D. J. Hoysak, K. J. Metz, and C. G. Eckert. 1991. A retrospective analysis of red-band effects on red-winged blackbirds. *Condor* 93:1013–1016.

Weatherhead, P. J., K. J. Metz, G. F. Bennett, and R. E. Irwin. 1993. Parasite faunas, testosterone and secondary sexual traits in male red-winged blackbirds. *Behav. Ecol. Sociobiol.* 33:13–24.

Weatherhead, P. J., and R. J. Robertson. 1977a. Harem size, territory quality, and reproductive success in the redwinged blackbird (*Agelaius phoeniceus*). *Can. J. Zool.* 55:1261–1267.

Weatherhead, P. J., and R. J. Robertson. 1977b. Male behavior and female recruitment in the red-winged blackbird. *Wilson Bull.* 89:583–592.

Weatherhead, P. J., and R. J. Robertson. 1979. Offspring quality and the polygyny threshold: "the sexy son hypothesis." *Am. Nat.* 113:201–208.

Weatherhead, P. J., and K. L. Teather. 1991. Are skewed fledgling sex ratios in sexually dimorphic birds adaptive? *Am. Nat.* 138:1159–1172.

Webster, M. S. 1992. Sexual dimorphism, mating system and body size in New World blackbirds (Icterinae). *Evolution* 46:1621–1641.

Wesolowski, T. 1987. Polygyny in three temperate forest Passerines (with a critical reevaluation of hypotheses for the evolution of polygyny). *Acta Ornithol.* 23:273–302.

West-Eberhard, M. J. 1992. Adaptation: current usages. Pp. 13–18 in E. F. Keller and E. A. Lloyd, eds., *Keywords in Evolutionary Biology*. Harvard University Press, Cambridge, Mass.

Westneat, D. F. 1992a. Do female red-winged blackbirds engage in a mixed mating strategy? *Ethology* 92:7–28.

Westneat, D. F. 1992b. Nesting synchrony by female red-winged blackbirds: effects on predation and breeding success. *Ecology* 73:2284–2294.

Westneat, D. F. 1993a. Polygyny and extra-pair fertilizations in eastern red-winged blackbirds (*Agelaius phoeniceus*). *Behav. Ecol.* 4:49–60.

Westneat, D. F. 1993b. Temporal patterns of within-pair copulation, male mate guarding, and extra-pair events in eastern red-winged blackbirds (*Agelaius phoeniceus*). *Behaviour* 124:291–312.

Westneat, D. F., and P. W. Sherman. 1993. Parentage and the evolution of parental behavior. *Behav. Ecol.* 4:66–77.

Westneat, D. F., P. W. Sherman, and M. L. Morton. 1990. The ecology and evolution of extra-pair copulations in birds. *Current Ornithol.* 7:331–369.

Wheelwright, N. T., C. B. Schultz, and P. J. Hodum. 1992. Polygyny and male parental care in savannah sparrows: effects on female fitness. *Behav. Ecol. Sociobiol.* 31:279–289.

Whittingham, L. A. 1989. An experimental study of paternal behavior in red-winged blackbirds. *Behav. Ecol. Sociobiol.* 25:73–80.

Whittingham, L. A. 1994. Additional mating opportunities and male parental care in red-winged blackbirds: a female removal experiment. *Anim. Behav.* in press.

Whittingham, L. A., A. Kirkconnell, and L. M. Ratcliffe. 1992a. Differences in song and sexual dimorphism between Cuban and North American red-winged blackbirds (*Agelaius phoeniceus*). *Auk* 109:928–933.

Whittingham, L. A., and R. J. Robertson. 1993. Nestling hunger and parental care in red-winged blackbirds. *Auk* 110:240–246.

Whittingham, L. A., and R. J. Robertson. 1994. The effect of food availability on the reproductive behaviour of red-winged blackbirds. *J. Anim. Ecol.* in press.

Whittingham, L. A., P. D. Taylor, and R. J. Robertson. 1992b. Confidence of paternity and male parental care. *Am. Nat.* 139:1115–1125.

Wiens, J. A. 1965. Behavioral interactions of red-winged blackbirds and common grackles on a common breeding ground. *Auk* 82:356–374.

Wiley, R. H. 1974. Evolution of social organization and life-history patterns among grouse. *Quart. Rev. Biol.* 49:201–227.

Wiley, R. H., and S. A. Harnett. 1976. Effects of interactions with older males on behavior and reproductive development in first-year male red-winged blackbirds (*Agelaius phoeniceus*). *J. Exp. Zool.* 196:231–242.

Wiley, R. H., and M. S. Wiley. 1980. Spacing and timing in the nesting ecology of a tropical blackbird: comparison of populations in different environments. *Ecol. Monogr.* 50:153–178.

Williams, G. C. 1966. *Adaptation and Natural Selection.* Princeton University Press, Princeton, N. J.

Williams, G. C. 1975. *Sex and Evolution.* Princeton University Press, Princeton, N. J.

Williams, L. 1952. Breeding behavior of the Brewer blackbird. *Condor* 54:3–47.

Willson, M. F. 1966. Breeding ecology of the yellow-headed blackbird. *Ecol. Monogr.* 36:51–77.

Wilson, E. O. 1975. *Sociobiology: The New Synthesis.* Harvard University Press, Cambridge, Mass.

Wimberger, P. H. 1988. Food supplement effects on breeding time and harem size in the red-winged blackbird (*Agelaius phoeniceus*). *Auk* 105:799–802.

Wingfield, J. C. 1988. The challenge hypothesis: interrelationships of testosterone and behavior. *Proc. XIX Intern. Ornithol. Congr.* 2:1685–1691.

Wingfield, J. C., G. F. Ball, A. M. Dufty, Jr., R. E. Hegner, and M. Ramenofsky. 1987. Testosterone and aggression in birds. *Am. Sci.* 75:602–608.

Winkler, D. W. 1987. A general model for parental care. *Am. Nat.* 130:526–543.

Wittenberger, J. F. 1976. The ecological factors selecting for polygyny in altricial birds. *Am. Nat.* 110:779–799.

Wittenberger, J. F. 1978. The breeding biology of an isolated bobolink population in Oregon. *Condor* 80:355–371.

Wittenberger, J. F. 1979. The evolution of mating systems in birds and mammals. Pp. 271–349 in P. Marler and J. Vandenbergh, eds., *Social Behaviour and Communication.* Plenum, New York.

Wittenberger, J. F. 1981. *Animal Social Behavior.* Duxbury, Boston.

Wittenberger, J. F. 1982. Factors affecting how male and female bobolinks apportion parental investments. *Condor* 84:22–39.

Wolf, L., E. D. Ketterson, and V. Nolan, Jr. 1988. Paternal influence on growth and survival of dark-eyed junco young: do parental males benefit? *Anim. Behav.* 36:1601–1618.

Wright, P. L., and M. H. Wright. 1944. The reproductive cycle of the male red-winged blackbird. *Condor* 46:46–59.

Yasukawa, K. 1978. Aggressive tendencies and levels of a graded display: factor analysis of response to song playback in the redwinged blackbird (*Agelaius phoeniceus*). *Behav. Biol.* 23:446–459.

Yasukawa, K. 1979. Territory establishment in red-winged blackbirds: importance of aggressive behavior and experience. *Condor* 81:258–264.

Yasukawa, K. 1981a. Male quality and female choice of mate in the red-winged blackbird (*Agelaius phoeniceus*). *Ecology* 62:922–929.

Yasukawa, K. 1981b. Song repertoires in the red-winged blackbird (*Agelaius phoeniceus*): a test of the Beau Geste hypothesis. *Anim. Behav.* 29:114–125.

Yasukawa, K. 1981c. Song and territory defense in the red-winged blackbird. *Auk* 98:185–187.

Yasukawa, K. 1987. Breeding and nonbreeding season mortality of territorial male red-winged blackbirds (*Agelaius phoeniceus*). *Auk* 104:56–62.

Yasukawa, K. 1989. The costs and benefits of a vocal signal: the nest-associated 'Chit' of the female red-winged blackbird, *Agelaius phoeniceus*. *Anim. Behav.* 38:866–874.

Yasukawa, K. 1990. Does the "Teer" vocalization deter prospecting female red-winged blackbirds? *Behav. Ecol. Sociobiol.* 26:421–426.

Yasukawa, K., J. L. Blank, and C. B. Patterson. 1980. Song repertoires and sexual selection in the red-winged blackbird. *Behav. Ecol. Sociobiol.* 7:233–238.

Yasukawa, K., R. A. Boley, J. L. McClure, and J. Zanocco. 1992a. Nest dispersion in the red-winged blackbird. *Condor* 94:775–777.

Yasukawa, K., R. A. Boley, S. E. Simon. 1987a. Seasonal change in the vocal behaviour of female red-winged blackbirds, *Agelaius phoeniceus*. *Anim. Behav.* 35:1416–1423.

Yasukawa, K., R. L. Knight, and S. K. Skagen. 1987b. Is courtship intensity a signal of male parental care in red-winged blackbirds (*Agelaius phoeniceus*)? *Auk* 104:628–634.

Yasukawa, K., F. Leanza, and C. D. King. 1993. An observational and brood-exchange study of paternal provisioning in the red-winged blackbird, *Agelaius phoeniceus*. *Behav. Ecol.* 4:78–82.

Yasukawa, K., J. L. McClure, R. A. Boley, and J. Zanocco. 1990. Provisioning of nestlings by male and female red-winged blackbirds, *Agelaius phoeniceus*. *Anim. Behav.* 40:153–166.

Yasukawa, K., and W. A. Searcy. 1981. Nesting synchrony and dispersion in red-winged blackbirds: is the harem competitive or cooperative? *Auk* 98:659–668.

Yasukawa, K., and W. A. Searcy. 1982. Aggression in female red-winged blackbirds: a strategy to ensure male parental investment. *Behav. Ecol. Sociobiol.* 11:13–17.

Yasukawa, K., and W. A. Searcy. 1985. Song repertoires and density assessment in red-winged blackbirds: further tests of the Beau Geste hypothesis. *Behav. Ecol. Sociobiol.* 16:171–175.

Yasukawa, K., and W. A. Searcy. 1986. Simulation models of female choice in red-winged blackbirds. *Am. Nat.* 128:307–318.

Yasukawa, K., L. K. Whittenberger, and T. A. Nielsen. 1992b. Anti-predator vigilance in the red-winged blackbird, *Agelaius phoeniceus*: do males act as sentinels? *Anim. Behav.* 43:961–969.

Young, H. 1963. Age-specific mortality in the eggs and nestlings of blackbirds. *Auk* 80:145–155.

Zimmerman, J. L. 1966. Polygyny in the dickcissel. *Auk* 83:534–546.

Author Index

Taxonomic Index

Subject Index